DISCRETE MATHEMATICS FOR COMPUTING

DISCRETE MATHEMATICS FOR COMPUTING

Peter Grossman
Department of Mathematics
Monash University, Australia

First published 1995 by
MACMILLAN PRESS LTD
Houndmills, Basingstoke, Hampshire RG21 2XS
and London
Companies and representatives
throughout the world

ISBN 0 333 64694 0 paperback

A catalogue record of this book is available from the British Library.

10   9   8   7   6   5   4   3
04   03   02   01   00   99

Printed in Hong Kong

CONTENTS

LIST OF SYMBOLS

Logic

$p \wedge q$	p and q	4.3
$p \vee q$	p or q	4.3
$\neg p$	not p	4.3
$p \rightarrow q$	if p then q	4.3
$p \leftrightarrow q$	p if and only if q	4.3
$p \oplus q$	p xor q (i.e. p or q but not both)	4: Exercise 3
$p \mid q$	p nand q (i.e. not both p and q)	4: Exercise 17
$P \equiv Q$	P is logically equivalent to Q	4.5 and 4.6
$\forall$	for all	4.7
$\exists$	there exists	4.7

Sets

$\{\ldots\}$	'the set of ...'	5.1
$:$	such that	5.1
$x \in A$	x is an element of A	5.1
$x \notin A$	x is not an element of A	5.1
$\varnothing$	null set	5.1
$\mathscr{E}$	universal set	5.1
$\mathbf{N}$	the set of natural numbers	5.1
$\mathbf{J}$	the set of integers	5.1
$\mathbf{Q}$	the set of rational numbers	5.1
$\mathbf{R}$	the set of real numbers	5.1
$A \subseteq B$	A is a subset of B	5.2
$A \cap B$	intersection of A and B	5.2
$A \cup B$	union of A and B	5.2
$\overline{A}$	complement of A	5.2
$A - B$	(set) difference of A and B	5.2
$|A|$	cardinality of A	5.3
$\mathscr{P}(A)$	power set of A	5.3
$A \times B$	Cartesian product of A and B	5.3
A^n	n-fold Cartesian product of A with itself	5.3
$(x_1, \ldots, x_n)$	ordered n-tuple	5.3
$E(x)$	equivalence class containing x	5.5

Functions

$f: A \to B$	f is a function with domain A and codomain B	6.1
$f(x)$	image of x under f	6.1
$g \circ f$	composite function of f and g	6.2
i	identity function (on a set)	6.2
f^{-1}	inverse of f	6.2

Boolean algebra

$x + y$	Boolean sum of x and y	8.1
$x \times y$	Boolean product of x and y	8.1
x'	Boolean complement of x	8.1
0	Boolean zero	8.1
1	Boolean one	8.1
	OR gate	8.3
	AND gate	8.3
	inverter (or NOT gate)	8.3

Combinatorics

$n!$	n factorial	9.3
$^{n}P_{r}$	number of permutations of n objects chosen from a set of r objects	9.3
$^{n}C_{r}, \binom{n}{r}$	number of combinations of n objects chosen from a set of r objects	9.4

Graphs

$\deg(v)$	degree of vertex v	10.2
$\overline{G}$	complement of G	10.2
$d(u, v)$	distance from u to v	11.3
$weight(e)$	weight of edge e	11.2 and 11.3

Number theory

$a \mid b$	a divides b	12.2
$a \nmid b$	a does not divide b	12.2
$\gcd(a, b)$	greatest common divisor of a and b	12.3
$\mathrm{lcm}(a, b)$	least common multiple of a and b	12.3
$a \equiv b \bmod m$	a is congruent to b modulo m	12.4

Other symbols

{...}	comment (in an algorithm)	1.1
←	'is assigned the value of' (in an algorithm)	1.1
a div b	integer quotient of a and b	2.3
a mod b	remainder after a is divided by b	2.3
$\lfloor n \rfloor$	integer part of n	2.3
frac(n)	fractional part of n	2.3
$\sum_{i=m}^{n}$	sum with i varying from m to n	7.2 and 7: Exercise 13
O(f)	'big O' of f	13.4
lg n	logarithm to the base 2 of n	13.5

PREFACE

This book is an introductory text on a number of topics in discrete mathematics, intended primarily for students undertaking a first degree in computing. The book has grown out of a set of my lecture notes used for the past two years in the first-year one-semester subject Discrete Mathematics at the Caulfield and Peninsula campuses of Monash University. The subject is available to students who are undertaking the Bachelor of Computing degree with a major in computer technology, information systems, software development, or computer application development.

Over the past 15 years or so, there has been an enormous expansion in the number and variety of degree courses being offered in computing and computer-related areas. Over a similar period, the term 'discrete mathematics' has been used widely to describe the collection of topics that form the prerequisite mathematical knowledge for studies in computing. Many textbooks are available with the words 'discrete mathematics' and either 'computing' or 'computer science' in their titles. These books generally cover the same broad range of topics: symbolic logic, sets, functions, induction, recursion, Boolean algebra, combinatorics, graph theory, and number theory, and also in some cases probability theory, abstract algebra, and mathematical models of computation. The unifying themes in these otherwise rather disparate topics are an emphasis on finite or countably infinite (hence 'discrete') mathematical structures, the use of an algorithmic approach to solving problems, and the applicability of the topics to problems arising in the study of computers and computing.

Academic authors often justify writing a new textbook by claiming that there was no book on the market that met the needs of their course. This is indeed the case here, but some further explanation is in order. There are many excellent discrete mathematics textbooks addressed principally to students of computer science. The reader of these books is typically assumed to have obtained at least a solid grounding in mathematics at final-year school level, the mathematics is often presented in a rather abstract style, and many of the examples used to illustrate the theory are drawn from computer science.

The students entering the Bachelor of Computing degree at Monash, however, come from a wider range of mathematical backgrounds. Some have substantial mathematical studies behind them, but there are others who have studied only a little mathematics in their final two years of school, in some cases many years before they began their tertiary studies. Furthermore, the course has a strong emphasis on information systems and commercial computing, and so the more abstract topics in theoretical computer science hold little interest for these students.

My aim, therefore, in writing this book, has been to cover the topics in discrete mathematics that I believe are appropriate for a course of this nature, and to present them in a way that makes them accessible to students with a modest mathematical background. A knowledge of calculus is not required, and even the level of facility assumed in algebra is fairly basic. A student who has seen sets and functions before will be at an advantage, but these topics are not prerequisite knowledge. A knowledge of programming is not required either, although a student who has done some programming will be able to relate that work to the material on algorithms in this book.

A brief summary of the contents of the book follows. Algorithms are discussed in Chapter 1, and a pseudocode notation is established, which is then used frequently in the subsequent chapters. Chapter 2 deals with non-decimal number systems, and in Chapter 3 the representation of integers and real numbers in a computer is explained. This is perhaps an unusual topic to find in a discrete mathematics textbook, but it provides an opportunity for students to relate mathematics to their work in other subjects. Chapter 4 is an introduction to propositional and predicate logic. Chapter 5 deals with sets and relations, and Chapter 6 with functions. Given the difficulty that many students have with grasping abstract mathematical concepts, the treatment of these topics in particular is gradual and informal. In Chapter 6, the emphasis is on thinking of functions as processes with input and output; the 'ordered pair' definition of a function is not mentioned. An unusual feature in Chapter 6 is a discussion of the relationship between the mathematical and programming concepts of a function. Chapter 7 deals with recursive algorithms and proof by induction, and includes a brief section on recursion in programming languages. Chapter 8 begins with a careful introduction to Boolean algebra, in which it is made clear that this topic is not just propositional logic in another notation but a branch of mathematics that can be used to unify the study of logic, sets, and other structures. The sections on digital circuit design and Karnaugh maps may seem out of place in a textbook that claims to have an emphasis on information systems, but these topics provide a realistic example of an area in which Boolean algebra is used, and for this reason I believe that a worthwhile purpose is served by including them. Chapter 9 deals with basic combinatorics. Graph theory is introduced in Chapter 10, and this leads into the study of trees and their applications in Chapter 11. Chapter 12 is an introduction to number theory, and includes a section on the application of modular arithmetic to the generation of pseudo-random numbers. Finally, Chapter 13 provides an introduction to the concepts of the time complexity and tractability of algorithms.

The exercises at the end of each chapter include questions that provide practice in routine computational techniques, as well as some more challenging problems designed to develop a deeper level of understanding of the material. Answers to all exercises for which a short answer can sensibly be given are provided at the end of the book.

My aim throughout the book has been to present the material in a fairly informal style, without sacrificing mathematical correctness. I am aware of the danger that, in aiming for informality, one can easily write mathematical statements that do not stand up to scrutiny when analysed carefully. It would be very easy, for example, to inadvertently define a cycle in a graph in a way that admits a single vertex as a cycle, which of course would then make nonsense of the definition of a tree as a connected graph with no cycles. The definitions in this book may not always be expressed with the level of mathematical formality to which professional mathematicians are accustomed, but they do mean what they say!

I would like to express my gratitude to a number of people whose assistance has been indispensable in the preparation of this book. First, I wish to thank my colleagues at Monash University: Cristina Varsavsky and Pam Norton from the Department of Mathematics, Ian Kirkwood from the Syme Division of Econometrics, and Chris Avram from the Department of Computer Technology, whose criticisms of a draft version of the manuscript were invaluable to me when I was putting the text into its final form. I am very grateful to the students taking the subject Discrete Mathematics who made suggestions for improving the book and pointed out some errors. The comments made by the anonymous reviewers appointed by Macmillan were also helpful, and were greatly appreciated. The artwork for the diagrams was very capably produced by Liz Butler and Jean Sheldon, to whom I express my sincere thanks. My thanks go to Marjorie Pressley for her careful and thorough editing. I acknowledge the support of the Department of Mathematics at Monash University in providing me with the time and facilities I needed while I was writing the book. Finally, I would like to thank Peter Debus, Elizabeth Gibson, and the staff at Macmillan Education Australia for their assistance during the production of this book.

1. INTRODUCTION TO ALGORITHMS

1.1 *What is an algorithm?*

A central theme in computing is the design of a process for carrying out a task. The task might be sorting names into alphabetical order, finding the cheapest way to link a set of computing sites into a network, converting a number to its representation in the binary system, encrypting a message for secure transmission, designing a digital circuit for a microchip, or determining the shortest path to be followed by a robotic arm. There will be many occasions throughout this book when we will find ourselves investigating problems of this nature, and asking: How can this task be performed by a person or a computer? In each case, the answer will take the form of a precise sequence of steps known as an *algorithm*.

An everyday example of an algorithm is provided by the steps you carry out when you withdraw cash from an automatic teller machine. We could write them out like this:

1. Insert your card into the slot.
2. Key in your personal identification number (PIN).
3. Select 'Withdraw cash' from the menu.
4. Enter the amount you want to withdraw.
5. Take your card, cash and transaction slip.

Steps 1 to 5 form a sequence of instructions to be carried out one after another. Each instruction is clear and unambiguous. (To some extent, this is a matter of opinion; for example, Step 1 does not specify which way round to insert the card, and Steps 2 to 4 do not tell us what to do if the machine rejects any of the input. Nevertheless, most people would probably agree that the instructions are clear enough for a person of normal intelligence to follow.) Finally, and importantly, the process is guaranteed to stop after a finite number of steps (five in this case). These are the essential features of an algorithm.

Definition

An *algorithm* is a finite sequence of steps for performing a task, such that:

- each step is a clear and unambiguous instruction, which can be executed in a finite time;
- the sequence in which the steps are to be executed is clearly defined;
- the process is guaranteed to stop after a finite number of steps have been executed.

Notice that the algorithm in our example has some other properties that are not included in the definition, but ones that we would expect any useful algorithm to have. First, there is *input* (in this case the information keyed in), and *output* (the cash and the transaction slip). Second, the algorithm has been designed with a *purpose* in mind (to withdraw cash).

The definition of an algorithm may remind you of a computer program. A *program* can be regarded as the implementation of an algorithm in a particular programming language. We use the term 'programming language' here to include not only 'traditional' programming languages such as COBOL, Pascal, C and Lisp, but also the macro command languages used in spreadsheets (such as Lotus 1-2-3®) and database management software (such as dBASE®). Designing an algorithm is one of the steps in writing a program. For the rest of this book, we will be primarily concerned with algorithms that are intended for implementation as programs on a computer.

In order to separate the process of designing an algorithm from the other aspects of programming in a particular language, the algorithms in this book will be written in a form of structured English called *pseudocode*. This will allow us to concentrate on the structure of the algorithm itself, without getting sidetracked by the details of a particular programming language.

Here is a simple example of an algorithm of the type that could be readily turned into a computer program.

EXAMPLE 1.1.1

Write an algorithm to calculate the area of a circle, given the radius.

SOLUTION

The area of a circle with radius r is πr^2. The algorithm is as follows:

1. Input r {r is the radius of the circle.}
2. $area \leftarrow \pi r^2$
3. Output $area$

This example illustrates several basic points about the way in which algorithms are written in pseudocode. The steps are numbered consecutively for easy reference. Explanatory comments that are not part of the algorithm itself are written between braces { }. The symbol ← denotes *assignment*; thus in Step 2, the formula πr^2 is evaluated, and the result is assigned to the variable *area*. (In computing terms, the result is stored in the memory in a location specified by the identifier *area*.)

Notice also that mathematical formulae such as πr^2 are written in the usual mathematical notation. When we write pseudocode we are interested only in the structure of the algorithm; we don't want to be concerned with the details of how a formula would be written in a programming language. (In Pascal, for example, Step 2 would appear as `area := pi * sqr(r)` in the program. Another Pascal statement would be needed to assign the appropriate numerical value to pi. These kinds of details vary from one language to another.)

1.2 *Control structures*

In Example 1.1.1, the steps are simply executed in order, one after the other. However, most algorithms contain one or more *control structures* — instructions that specify how many times other instructions are to be executed, or under what conditions they are to be executed. The next example illustrates a situation in which control structures are needed.

EXAMPLE 1.2.1

Find the smallest number in a list of numbers.

SOLUTION

The smallest number can be found by looking at each number in turn, keeping track at each step of the smallest number so far.

1. Input the number of values n
2. Input the list of numbers $x_1, x_2, \ldots, x_n$
3. *min* ← x_1
4. **For** $i = 2$ **to** n **do**
   4.1. **If** $x_i <$ *min* **then**
      4.1.1. *min* ← x_i
5. Output *min*

Two control structures appear in this example. The **For-do** in Step 4 causes the following step (or steps) to be executed a specified number of times. In this example, the *index variable*, i, ranges through the values 2, 3, 4, ..., n, and Step 4.1 is executed with i set equal to each of these values in turn. This type of structure is often referred to as a *loop*. The **If-then** in Step 4.1 is another control structure — Step 4.1.1 is executed if $x_i < min$, otherwise it is ignored.

Notice how the logical structure of the algorithm is indicated by the use of indenting, and given further emphasis by the way the steps are numbered — Step 4.1.1 is part of Step 4.1, which in turn is part of Step 4.

A list of some commonly used control structures is given in Table 1.1. They are available in many programming languages.

Control structure	Example of use
If-then	1. **If** $x < 0$ **then** 1.1. $x \leftarrow -x$

If-then-else	1. **If** $x \geq 0$ **then**
	      1.1. $y \leftarrow \sqrt{x}$
	   else
	      1.2. Output '$\sqrt{x}$ does not exist.'

For-do	1. $sum \leftarrow 0$ 2. **For** $i = 1$ **to** 10 **do** 2.1. $sum \leftarrow sum + i^2$

While-do	1. **While** $answer \neq$ 'y' and $answer \neq$ 'n' **do**
	      1.1. Output 'Please answer y or n.'
	      1.2. Input $answer$

Repeat-until	1. $i \leftarrow 0$
	2. **Repeat**
	      2.1. $i \leftarrow i + 1$
	      2.2. Input x_i
	   until $x_i = 0$

Table 1.1

When once we have designed an algorithm to perform a particular task, we would naturally like to find out whether it does the task correctly. One useful technique, known as *tracing* (or desk-checking) an algorithm, is carried out by choosing a particular set of inputs and recording the values of all the variables at each step of the algorithm.

EXAMPLE 1.2.2

Trace the algorithm in Example 1.2.1 with inputs $n = 3$, $x_1 = 5$, $x_2 = 4$, $x_3 = 8$.

SOLUTION

See Table 1.2.

Step	*min*	*i*	x_i	Output
3	5	–	–	–
4	5	2	4	–
4.1	5	2	4	–
4.1.1	4	2	4	–
4	4	3	8	–
4.1	4	3	8	–
5	4	3	8	4

Table 1.2

Since 4 is the smallest number in the list 5, 4, 8, the trace table confirms that the algorithm gives the correct output for this set of inputs.

Of course, we cannot claim on the strength of the trace in Example 1.2.2 that the algorithm is correct for every possible set of inputs. A trace can reveal that an algorithm is wrong, but it can never be used to prove that an algorithm is correct.

The next example illustrates a situation where a **While-do** or a **Repeat-until** is useful.

EXAMPLE 1.2.3

Design an algorithm to check whether a string $c_1c_2...c_n$ of n characters consists entirely of digits or whether non-digit characters are present, and output an appropriate message.

SOLUTION

We could use a **For-do** loop to check every character in the string, but it would be more efficient to stop checking as soon as a non-digit character is encountered. One way of doing this is to use a **Repeat-until**, with a counter i to keep track of the position in the string:

1.[1] $i \leftarrow 1$; *nondigit_detected* $\leftarrow$ false
2. **Repeat**

---

[1]Step 1 is really two steps, written on one line for convenience and separated by a semicolon. This is often done where several variables are to be initialised (given initial values), as is the case here.

2.1. **If** c_i is not a digit **then**
    2.1.1. *nondigit_detected* ← true
2.2. $i \leftarrow i + 1$
until *nondigit_detected* = true

We will always write identifiers in algorithms as strings of characters without spaces, using the underscore character where necessary if an identifier is made of two or more English words (as we have done with *nondigit_detected* above).

The variable *nondigit_detected* is an example of a *logical* (or Boolean) variable; it may take only the values 'true' and 'false'. Logical variables are often useful in control structures such as the **Repeat-until** in this example. (It would be permissible to omit '= true' from the last line, and just write '**until** *nondigit_detected*'.)

The counter i in this algorithm works in much the same way as the index i of the **For-do** in Example 1.2.1; it is initialised to 1 before the loop is entered, and incremented by 1 at the end of each pass through the loop. An important difference is that the initialisation and incrementation are built into the structure of the **For-do**, and so it is not necessary to write the step $i \leftarrow i + 1$ explicitly there; in fact, it would be an error to do so.

The process we have just described will stop with *nondigit_detected* = true as soon as a non-digit is detected, but it fails if the string contains only digits. We can fix this by testing at the end of each pass through the loop whether the end of the string has been reached. After the last character c_n has been checked, i is incremented in Step 2.2 to $n + 1$, so this is what we need to test for:

1. $i \leftarrow 1$; *nondigit_detected* ← false
2. **Repeat**
    2.1. **If** c_i is not a digit **then**
        2.1.1. *nondigit_detected* ← true
    2.2. $i \leftarrow i + 1$
    until *nondigit_detected* = true or $i = n + 1$

This looks better; if there are no non-digit characters, the loop is executed n times and finishes with *nondigit_detected* = false. It remains only to add suitable input and output steps to complete the algorithm:

1. Input n
2. Input $c_1 c_2 \ldots c_n$
3. $i \leftarrow 1$; *nondigit_detected* ← false
4. **Repeat**
    4.1. **If** c_i is not a digit **then**
        4.1.1. *nondigit_detected* ← true
    4.2. $i \leftarrow i + 1$
    until *nondigit_detected* = true or $i = n + 1$
5. **If** *nondigit_detected* = true **then**

      5.1. Output 'The string contains non-digit characters.'
      else
      5.2. Output 'The string consists entirely of digits.'

What happens if a null string ($n = 0$) is input? In this case we would be in trouble, as the loop would be entered with $i = 1$, and Step 4.1 could not be executed. This illustrates a limitation of the **Repeat-until** construct — a loop of this type *always executes at least once*.

    One way to avoid the problem is to use a **While-do**; because the test is performed at the beginning of the loop rather than at the end, it is possible for the steps within a **While-do** loop not to be executed at all. Here is the algorithm again, this time rewritten using a **While-do**:

1. Input n
2. Input $c_1 c_2 \dots c_n$
3. $i \leftarrow 0$; *nondigit_detected* $\leftarrow$ false
4. **While** *nondigit_detected* = false and $i < n$ **do**
    4.1. $i \leftarrow i + 1$
    4.2. **If** c_i is not a digit **then**
        4.2.1. *nondigit_detected* $\leftarrow$ true
5. **If** *nondigit_detected* = true **then**
    5.1. Output 'The string contains non-digit characters.'
    else
    5.2. Output 'The string consists entirely of digits.'

This algorithm works even if $n = 0$.

1.3 *Further examples of algorithms*

In this section, we present three more examples showing how algorithms are developed, beginning with the specification of a problem, and following through the steps to the final algorithm.

EXAMPLE 1.3.1

Design an algorithm to evaluate x^n, where x is any real number and n is a positive integer. (Assume that the operation of multiplication is available, but raising to a power is not; this is the case, for example, in the programming language Pascal.)

SOLUTION

If n is a positive integer, x^n can be evaluated using the formula:

$$x^n = \underbrace{x \times x \times \dots \times x}_{n \text{ times}}$$

In order to carry out this process, we will need a variable *answer*, which is initially assigned the value of x, and which is then multiplied by x the required number of times. The number of multiplications is fixed at $n - 1$, so a **For-do** loop is the most appropriate control structure to use here.

1. Input x, n
2. *answer* $\leftarrow x$
3. **For** $i = 1$ **to** $n - 1$ **do**
    3.1. *answer* $\leftarrow$ *answer* $\times x$
4. Output *answer*

Is this sequence of steps an algorithm? The steps are clear and unambiguous, they are executed in a clearly defined sequence, and Step 3.1 in the **For-do** loop is executed a finite number of times, so the process is guaranteed to terminate eventually. Therefore all the requirements of an algorithm are satisfied.

Is the algorithm correct? Table 1.3 is a trace table with inputs $x = 2$ and $n = 3$:

Step	i	*answer*	Output
2	–	2	–
3	1	2	–
3.1	1	4	–
3	2	4	–
3.1	2	8	–
4	2	8	8

Table 1.3

The output is correct in this case, since $2^3 = 8$.

It is also a good idea to check the algorithm using input values that lie at the extremes of the allowed range of inputs, because in practice this is often where an algorithm will fail. In this example, n must be greater than or equal to 1, so we should check that the algorithm gives the correct output when $n = 1$. If $n = 1$ then Step 3 will read **For** $i = 1$ **to** 0 **do** ..., which means that Step 3.1 is not executed at all. (In general, the steps within a **For-do** are not executed if the final value of the loop index is less than the initial value.) The output is therefore simply the original value of *answer*, namely x. This is exactly what it should be, because $x^1 = x$.

With a simple algorithm such as this one, the testing we have done is enough for us to be reasonably confident that it is correct (although of course we have not proved that it is). More extensive testing is needed if the algorithm is more complex.

Before we leave this example, let's consider a slightly different version of the problem. Suppose we were asked to design an algorithm to evaluate x^n, where x is any real number and n is a *non-negative* integer. Could we use the same algorithm? In order to answer this question, we need to ascertain whether the algorithm works if $n = 0$ is input. If $n = 0$, Step 3.1 is not executed, and the value of x is output. This is not the correct output, because $x^0 = 1$, so we conclude that the algorithm will need to be altered. You are asked to do this in Exercise 7 of this chapter.

EXAMPLE 1.3.2

Design an algorithm to swap the values of two variables. (In practice, such an algorithm might form part of a larger algorithm for sorting data into numerical or alphabetical order.)

SOLUTION

Let the two variables be x and y. We could try something like this:

1. $x \leftarrow y$
2. $y \leftarrow x$

This looks quite plausible at first, but a trace quickly reveals that something is wrong!

Table 1.4 shows a trace with $x = 2$ and $y = 3$:

Step	x	y
Initially	2	3
1	3	3
2	3	3

Table 1.4

The problem occurs because the original value of x is lost when Step 1 is executed. One way to prevent this from happening is to store the original value in another variable. This can be done in the following way:

1. $temp \leftarrow x$
2. $x \leftarrow y$
3. $y \leftarrow temp$

You should check that the trace now gives the correct result.

EXAMPLE 1.3.3

The applicants for a certain position are given an aptitude test consisting of 20 multiple-choice questions. Design an algorithm to output a list of the applicants (identified by number), their scores, and a message stating whether they are to be short-listed for the position (those who score 16 or more) or reconsidered at a later date (those who score in the range from 12 to 15). The input consists of the answers provided by each applicant.

SOLUTION

We will use *number_of_applicants* to denote the number of applicants. The same procedure will need to be applied to each applicant, so a **For-do** is the appropriate structure to use, with the index i ranging from 1 to *number_of_applicants*.

We denote the 20 answers submitted by Applicant i by $a_{i,1}$, $a_{i,2}$, ..., $a_{i,20}$. In order to calculate the applicant's score, the answer $a_{i,q}$ provided by Applicant i to Question q will need to be compared with the correct answer for that question, which we denote by c_q. Another **For-do** loop, 'nested' within the first one and with q ranging from 1 to 20, will be needed to accomplish this.

In the final steps of the outer **For-do**, the output for Applicant i will need to be generated according to the specifications in the problem.

1. Input *number_of_applicants*
2. **For** $i = 1$ **to** *number_of_applicants* **do**
    2.1. *score* $\leftarrow$ 0
    2.2. **For** $q = 1$ **to** 20 **do**
        2.2.1. Input $a_{i,q}$ {$a_{i,q}$ is Applicant i's answer to Question q}
        2.2.2. **If** $a_{i,q} = c_q$ **then** {c_q is the correct answer to Question q}
            2.2.2.1. *score* $\leftarrow$ *score* + 1
    2.3. Output i, *score*
    2.4. **If** *score* $\geq$ 16 **then**
        2.4.1. Output 'Short-listed'
        else if *score* $\geq$ 12 **then**
        2.4.2. Output 'Reconsider later'

The layout of Step 2.4 needs some explanation. Strictly speaking, according to the rules we have been following, the **if-then** after the **else** should be shown as a separate numbered statement, like this:

2.4. **If** *score* $\geq$ 16 **then**
    2.4.1. Output 'Short-listed'
    else
    2.4.2. **If** *score* $\geq$ 12 **then**
        2.4.2.1. Output 'Reconsider later'

In practice, however, we think of the conditions '*score* $\geq$ 16' and '*score* $\geq$ 12' as specifying two cases that require different actions to be taken. The layout we have used reflects this way of looking at it.

The purpose of this chapter has been to present a brief introduction to algorithms and to define the pseudocode we will be using. Algorithms play a central role in computing and discrete mathematics, and they will reappear frequently in subsequent chapters.

Exercises

1. Modify the algorithm in Example 1.2.1 so that the output also includes the position in the list where the smallest number occurs.

2. Write an algorithm to input a period of time in hours, minutes and seconds, and output the time in seconds.

3. Write an algorithm to calculate $1^2 + 2^2 + 3^2 + \ldots + n^2$, the sum of the first n perfect squares, and output the result.

4. Write an algorithm to input the price of a purchase and the amount tendered, and calculate the change due. An appropriate message should be output if the amount tendered is less than the purchase price.

5. Write an algorithm to calculate the tax payable on a given taxable income, according to the following rules:

Taxable income	Tax
$1–$5400	0
$5401–$20700	0 plus 20 cents for each $1 over $5400
$20701–$36000	$3060 plus 38 cents for each $1 over $20700
$36001–$50000	$8874 plus 46 cents for each $1 over $36000
$50001 and over	$15314 plus 47 cents for each $1 over $50000

6. Write an algorithm to input a list of numbers and test whether the numbers are in increasing order of magnitude, giving an appropriate message as output. The algorithm should be designed so that testing stops as soon as a definite answer is obtained.

7. Modify the algorithm in Example 1.3.1 so that it gives the correct output when $n = 0$ is input.

8. The following algorithm calculates a number known as the 'digital root' of a positive integer.

   1. Input a positive integer n
   2. $d \leftarrow$ number of digits in n
   3. **While** $d > 1$ **do**
      3.1. $n \leftarrow$ sum of the digits of n
      3.2. $d \leftarrow$ number of digits in n
   4. Output n

   (a) Trace the algorithm when 8678 is input.
   (b) List all the possible values that the output of the algorithm could take.

9. Consider the following sequence of steps:

1. Input a non-negative integer n
2. $i \leftarrow 0$
3. **While** n is even **do**
   3.1. $n \leftarrow n/2$
   3.2. $i \leftarrow i + 1$
4. Output i

   (a) What is the output when 12 is input?
   (b) What is the output when any odd number is input?
   (c) What happens when 0 is input?
   (d) Is this sequence of steps an algorithm?

10. Consider the following sequence of steps:

1. Input a positive integer n
2. *answer* $\leftarrow n$
3. **While** $n > 1$ **do**
   3.1. $n \leftarrow n - 1$
   3.2. *answer* $\leftarrow$ *answer* $\times n$
4. Output *answer*

   (a) Construct a trace table to show what happens when 4 is input.
   (b) Is this sequence of steps an algorithm? Give a reason for your answer.

11. Write an algorithm to input a string of characters and test whether the parentheses (round brackets) in the string are paired correctly. (Use a variable *excess_left*, which records the excess of the number of left parentheses over the number of right parentheses as the algorithm looks at each character in turn. If *excess_left* is never negative, and the end of the string is reached with *excess_left* $= 0$, then the parentheses are paired correctly.)

12. Consider the following algorithm:

1. Input a positive integer n
2. **For** $i = 1$ **to** n **do**
   2.1. $a_i \leftarrow 0$
3. **For** $i = 1$ **to** n **do**
   3.1. **For** $j = 1$ **to** n **do**
      3.1.1. **If** j is divisible by i **then**
         3.1.1.1. $a_j \leftarrow 1 - a_j$
         {a_j is always either 0 or 1}
4. **For** $i = 1$ **to** n **do**
   4.1. Output a_i

   (a) List the values taken by $a_1, a_2, \ldots, a_n$ at the end of each pass through the outer **For-do** loop (Step 3) when the algorithm is run with $n = 10$.
   (b) Given any value of n, can you predict the final value of each of the a_is without tracing through the algorithm? Justify your answer.

13. Consider the process defined by the following sequence of steps:

    1. Input a positive integer n

    2. **While** $n \neq 1$ **do**

        2.1. **If** n is even **then**

            2.1.1. $n \leftarrow n/2$

        else

            2.1.2. $n \leftarrow 3n + 1$

    3. Output 'Finished!'

(a) List the successive values taken by n if an initial value of 7 is input.

(b) Explore the process for various other inputs.

(The problem of determining whether the process terminates for all possible inputs, and hence whether or not this sequence of steps is an algorithm, is unsolved; see the article 'On the ups and downs of hailstone numbers' by B. Hayes in the Computer Recreations column of *Scientific American*, Vol. 250, No. 1, January 1984, pp. 13–17.)

2. BASES AND NUMBER REPRESENTATION

2.1 Real numbers and the decimal number system

Numbers can be represented in various ways. For example, the number 'four' can be written in the usual notation as 4, in Roman numerals as IV, or even just as four tally marks, ||||. Similarly, 'two-and-a-half' can be written as the mixed number $2\frac{1}{2}$, as the improper fraction $\frac{5}{2}$, or as the decimal number 2.5. In particular, numbers can be represented using systems similar to the familiar decimal system but based on numbers other than 10. In this chapter, we investigate the representation of numbers using different number bases, paying particular attention to the number systems used in computing.

Before we look at other number bases, it is helpful to recall what types of numbers there are, and how they are represented in the decimal system:

- The *natural* numbers (also called the positive integers) are the numbers 1, 2, 3, 4, ...
- The *integers*, or whole numbers, include zero and the negative whole numbers as well as the natural numbers: ..., –3, –2, –1, 0, 1, 2, 3, ...
- The *rational* numbers are all the numbers that can be expressed in the form m/n where m and n are integers and n is not zero. Note that this includes all of the integers, as well as all 'fractions', both proper (e.g. $\frac{1}{2}$) and improper (e.g. $\frac{5}{3}$). Any rational number can also be written as a terminating or recurring decimal, for example $\frac{1}{4} = 0.25$, $\frac{1}{6} = 0.16666...$

One of the more surprising mathematical facts is that some of the numbers that we would like to be able to define and use are not rational numbers. For example, although $\sqrt{2}$ can be approximated by rational numbers to any degree of accuracy we choose, it is impossible to find integers m and n such that $m/n = \sqrt{2}$ exactly. We express this fact by saying that $\sqrt{2}$ is an *irrational* number. A proof that $\sqrt{2}$ is irrational is outlined in Exercise 15 at the end of this chapter. All roots (square roots, cube roots, etc.) of the natural numbers are irrational numbers, except those roots that are themselves natural numbers such as $\sqrt{4}$. The numbers π and e (the base of natural logarithms) are also irrational. The decimal representation of an irrational number does not terminate or recur, for example $\sqrt{2} = 1.4142135...$ and $\pi = 3.1415926...$

All the numbers we have introduced, both rational and irrational, are called *real* numbers. For example, 3, $\frac{7}{5}$, -4.38 and $2 - 3\sqrt{5}$ are all real numbers. (Unless you have studied complex numbers elsewhere, all of the numbers you have encountered in your studies are real numbers.)

A real number is written in the decimal system as a string of digits, preceded by a minus sign if the number is negative. The representation may also include a decimal point. (If no decimal point is written, there is an implied decimal point after the last digit.) Some real numbers, such as $\frac{1}{6}$ and $\sqrt{2}$, have non-terminating decimal representations; they cannot be represented exactly in this system using a finite number of digits.

The decimal system is an example of a *positional* number system, because each digit has a *place value* that depends on its position in relation to the decimal point. The digit immediately to the left of the point is the units (or ones) digit ($1 = 10^0$). To the left of the units digit is the tens digit ($10 = 10^1$), then the hundreds digit ($100 = 10^2$), the thousands digit ($1000 = 10^3$), and so on. Similarly, to the right of the decimal point are the tenths digit ($\frac{1}{10} = 10^{-1}$), the hundredths digit ($\frac{1}{100} = 10^{-2}$), and so on. The value of the number is obtained by multiplying each digit by its place value and adding the results.

For example, the decimal number 2386.75 can be expanded in the following way:

$$2386.75 = 2 \times 10^3 + 3 \times 10^2 + 8 \times 10^1 + 6 \times 10^0 + 7 \times 10^{-1} + 5 \times 10^{-2}$$

The decimal system is said to use a *base* of 10 because the place values are powers of 10.

2.2 *The binary number system*

It is likely that most of what we have said so far is already familiar to you. Because we use the decimal system all the time, we rarely give it a second thought, and it is easy to forget that there is no mathematical reason for using powers of 10 as the place values. The choice of 10 as the base presumably arose because people in ancient times used their 10 fingers for counting, but in fact any natural number greater than 1 can be used as the base of a positional number system.[1]

A familiar example of something like a non-decimal number system is the subdivision of an hour into 60 minutes and a minute into 60 seconds, giving a kind of base 60 number system. For example, a time of 2 hours 26 minutes and 35 seconds can be expressed in seconds as follows:

$$2 \text{ h } 26 \text{ m } 35 \text{ s} = 2 \times 60^2 + 26 \times 60^1 + 35 \times 60^0 \text{ seconds}$$

---

[1] Actually, the situation is more general than this — negative numbers, and even complex numbers, can be used as bases. An interesting discussion of the different number systems that can be constructed appears in Chapter 4 of *The Art of Computer Programming, Vol. 2: Seminumerical Algorithms*, 2nd ed., by D.E. Knuth (Addison–Wesley, 1981).

Notice the similarity of this expression to the expansion of a decimal number into powers of 10. (The analogy is not perfect, because there are no names for 60^3 seconds, 60^4 seconds, and so on.)

It turns out, for reasons that will be explained later in this chapter, that the bases 2, 8, and 16, corresponding to what are known as the binary, octal, and hexadecimal systems respectively, are particularly useful in computing.

The *binary* system is the positional number system that uses 2 as the base. Whereas the decimal system uses the 10 decimal digits 0, 1, 2, ..., 9, the binary system uses the 2 binary digits (or *bits*) 0 and 1. A positive number written in the binary system will appear as a string of zeros and ones, and may also include a point (don't call it a decimal point!); for example:

$$1101.01_2$$

The subscript 2 denotes the base. The base should always be indicated in this way in any work involving number systems with bases other than 10.

The place values of the digits in a binary number are powers of 2. Starting at the point and moving to the left, we have the units digit ($1 = 2^0$), the twos digit ($2 = 2^1$), the fours digit ($4 = 2^2$), the eights digit, and so on. To the right of the point are the halves digit, the quarters digit, the eighths digit, and so on. A binary number can be evaluated (and hence converted to decimal) by writing it in expanded form:

$$1101.01_2 = 1 \times 2^3 + 1 \times 2^2 + 0 \times 2^1 + 1 \times 2^0 + 0 \times 2^{-1} + 1 \times 2^{-2}$$
$$= 8 + 4 + 1 + 0.25$$
$$= 13.25$$

The reason that binary numbers arise so often in computing can be summarised as follows. In the digital computing systems in common use today, the devices that store data or information of any kind (whether magnetically on a disk or tape, electronically in the random access memory (RAM) in a microchip, or in some other way) consist of a large number of memory elements, each of which can be in one of two states (such as magnetised or unmagnetised, on or off). Similarly, when data is transmitted inside a computer or through a network, it is usually coded as a stream of signal elements that take one of two forms, such as the presence or absence of an electric current, or two alternating currents with different frequencies. (An exception is the modems used for data communication, which work with more than two states.) The manipulation of data to perform arithmetic and other computations takes place in the digital circuitry etched on a microchip, which also operates using currents of just two kinds. In short, the data handled by a digital computer is stored, transmitted and manipulated as a stream of information 'elements' of two types, which can be denoted by the symbols 0 and 1. Thus the binary number system is the most natural way of representing numbers in a digital computer.

Table 2.1 shows the integers from 0 to 20 in their binary and decimal representations.

Binary	Decimal	Binary	Decimal
0	0		
1	1	1011	11
10	2	1100	12
11	3	1101	13
100	4	1110	14
101	5	1111	15
110	6	10000	16
111	7	10001	17
1000	8	10010	18
1001	9	10011	19
1010	10	10100	20

Table 2.1

2.3 *Conversion from decimal to binary*

How can a number be converted from the decimal to the binary system? It should come as no surprise that the answer to this question will take the form of an algorithm. Different processes are used for converting the integer and fractional parts of a number, so we will consider the two cases separately, beginning with the integer part.

Examination of Table 2.1 reveals the following pattern:

- Odd numbers have a binary representation ending in 1.
- Even numbers have a binary representation ending in 0.

These two statements can be summarised as follows:

The last digit in the binary representation is the remainder after dividing the number by 2.

The first step in carrying out the conversion to binary is therefore to divide the decimal number by 2, to obtain a (whole number) *quotient* and a *remainder*. We will use the following (Pascal-style) notation to denote these operations:

n div 2 is the quotient when n is divided by 2

n mod 2 is the remainder when n is divided by 2

For example:

12 div 2 = 6, 12 mod 2 = 0

13 div 2 = 6, 13 mod 2 = 1

In order to see how to proceed further, we need to make another observation:

The binary number obtained by removing the rightmost bit from the binary representation of n is the binary representation of n div 2.

Therefore, in order to obtain the second bit from the right-hand end, we must perform another division by 2, starting with the quotient from the first division, and obtaining a new quotient and remainder. This process is repeated, the remainders at each step forming the digits of the answer from right to left, until a quotient of zero is obtained.

Here is the entire process written as an algorithm:

1. Input n {n must be a natural number}
2. **Repeat**
   2.1. Output n mod 2
   2.2. $n \leftarrow n$ div 2
   until $n = 0$

The outputs must be read in reverse order to yield the correct answer.

Table 2.2 shows a trace of the algorithm with $n = 6$:

Step	n	Output
1	6	–
2.1	6	0
2.2	3	–
2.1	3	1
2.2	1	–
2.1	1	1
2.2	0	–

Table 2.2

The answer, reading the output column from the bottom up, is 110_2.

In practice, it is more convenient to set out decimal-to-binary conversions in the manner shown in the next example.

EXAMPLE 2.3.1

Convert the decimal number 25 to its binary representation.

SOLUTION

The number to be converted (25) forms the first entry in a column, with the new base (2) written to the left. At each step, the last entry in the column is divided by 2, the quotient is written below the number being divided, and the remainder is written in another column to the right.

2	25	
	12	1
	6	0
	3	0
	1	1
	0	1

(Leaving out the last step, 1 div 2 = 0 and 1 mod 2 = 1, is a common mistake. The algorithm does not terminate until the quotient is zero.)

Reading the remainder column from the bottom up, we obtain the answer: $25_{10} = 11001_2$.

We now turn to the problem of converting a decimal fraction to binary. For example, suppose we want to find the binary representation of 0.375_{10}. Because 0.375 is less than 0.5, the first bit to the right of the point in the binary representation (the halves bit, which has a place value of $\frac{1}{2}$) will be 0. Expressed another way, the halves bit is 0 because 2×0.375 is less than 1. If, on the other hand, we had started with a number greater than or equal to $\frac{1}{2}$, say 0.875, the halves bit would be 1, and $2 \times 0.875 \geq 1$. We can sum up this observation in the following rule:

The halves bit in the binary representation of n is the integer part of 2n.

Some new notation will be needed in what follows:

$$\lfloor n \rfloor \text{ denotes the integer part of } n$$

$$\text{frac}(n) \text{ denotes the fractional part of } n$$

For example, $\lfloor 2.7 \rfloor = 2$ and $\text{frac}(2.7) = 0.7$.

The observation we have made suggests that the following process involving repeated multiplication by 2 can be used to convert a decimal fraction to its binary representation.

1. Input n
2. **Repeat**
   2.1. $m \leftarrow 2n$
   2.2. Output $\lfloor m \rfloor$
   2.3. $n \leftarrow \text{frac}(m)$
   until $n = 0$

There is one problem with this process — what happens if the condition $n = 0$ is never satisfied? This difficulty can certainly arise; we already know that the decimal representation of a fraction need not terminate (recall $\frac{1}{3} = 0.33333...$), and the same thing can happen with the binary representation. We conclude that this sequence of steps is not an algorithm, since it fails the requirement that an algorithm must terminate after a finite number of steps.

One way to avoid the difficulty is to specify on input the number of digits we want in the answer, and to output just that number of digits of the binary representation. Here is the process with the necessary changes made:

1. Input n, *digits*
2. $i \leftarrow 0$
3. **Repeat**
      3.1. $i \leftarrow i + 1$
      3.2. $m \leftarrow 2n$
      3.3. Output $\lfloor m \rfloor$
      3.4. $n \leftarrow \text{frac}(m)$
   until $n = 0$ or $i = digits$

In practice, the calculations are usually set out as shown in the next example.

EXAMPLE 2.3.2

Convert the decimal fraction 0.32_{10} to its binary representation, with 5 digits after the point.

SOLUTION

The number to be converted forms the first entry in a column, with the new base (2) written to the right. At each step, the last entry in the column is multiplied by 2, the fractional part is written below the number being multiplied, and the integer part is written in another column to the left. For convenience, the point in front of the fractional part is omitted.

	32	2
0	64	
1	28	
0	56	
1	12	
0	24	

The left column (read from the top down) gives the answer: $0.32_{10} = 0.01010..._2$. (Note that the answer is truncated to 5 digits after the point; it is not rounded off.)

If a number has both an integer part and a fractional part, simply convert each part separately and combine the results. For example, now that we have shown that $25_{10} = 11001_2$ and $0.32_{10} = 0.01010..._2$, it follows without any further work that $25.32_{10} = 11001.01010..._2$.

2.4 *The octal and hexadecimal systems*

The techniques described in the previous section can be generalised to bases other than 2. In particular, the bases 8 and 16 are often used in computing, for a reason that will be explained shortly.

In the base 8 or *octal* system, numbers are written using the 8 octal digits 0, 1, 2, 3, 4, 5, 6, 7. The place value of each digit is a power of 8.

For example:

$$374.2_8 = 3 \times 8^2 + 7 \times 8^1 + 4 \times 8^0 + 2 \times 8^{-1}$$
$$= 252.25_{10}$$

In the base 16 or *hexadecimal* system (often simply called 'hex'), 16 digits are needed. New symbols are required to denote the digits with values 10, 11, 12, 13, 14 and 15, and there is an established convention that the first six upper-case letters are used for this purpose. The 16 hexadecimal digits are therefore 0, 1, 2, ..., 9, A, B, C, D, E, F. The place value of each digit is a power of 16.

For example:

$$E9C.8_{16} = 14 \times 16^2 + 9 \times 16^1 + 12 \times 16^0 + 8 \times 16^{-1}$$
$$= 3740.5_{10}$$

Converting a decimal number to its octal or hexadecimal representation is similar to converting from decimal to binary; the only difference is that 8 or 16 is used in place of 2 as the divisor or multiplier.

EXAMPLE 2.4.1

Convert 275.4375_{10} to octal.

SOLUTION

Convert the integer part:

8	275	
	34	3
	4	2
	0	4

$$275_{10} = 423_8$$

Convert the fractional part:

	4375	8
3	5000	
4	0	

$$0.4375_{10} = 0.34_8$$

Combine the results to obtain the answer:

$$275.4375_{10} = 423.34_8$$

EXAMPLE 2.4.2

Convert 985.78125_{10} to hexadecimal.

SOLUTION

Convert the integer part:

16	985	
	61	9
	3	13
	0	3

$$985_{10} = 3D9_{16}$$

Convert the fractional part:

	78125	16
12	50000	
8	0	

$$0.78125_{10} = 0.C8_{16}$$

Combine the results to obtain the answer:

$$985.78125_{10} = 3D9.C8_{16}$$

We have seen why the binary system is important in computing; it is the system used internally by the computer itself. But why are the octal and hexadecimal systems useful?

The main drawback of using the binary system as a general purpose number system is that even moderately large integers have many digits in their binary representation (typically more than three times as many as in the decimal representation). The advantage of larger bases is that we can write numbers using fewer digits. However, the decimal system is inconvenient if we often have to convert between it and the binary system, because, as we have seen, the conversion can involve a substantial amount of calculation.

By using the octal and hexadecimal systems, we avoid the problem of large numbers of digits, while gaining an important advantage over the decimal system — there are simple algorithms for converting between binary and octal, and between binary and hexadecimal. For this reason, bases 2 and 8 are described as *related bases*. Similarly, bases 2 and 16 are related.

To convert a number from binary to octal, group the bits into sets of 3 on either side of the point. Each group of 3 bits corresponds to 1 digit in the octal representation.

EXAMPLE 2.4.3

| Convert 10100011.10111_2 to octal.

SOLUTION

$$\underbrace{10}_{2} \ \underbrace{100}_{4} \ \underbrace{011}_{3} \cdot \underbrace{101}_{5} \ \underbrace{110}_{6}$$

$$10100011.10111_2 = 243.56_8$$

Converting from octal to binary is equally easy — replace each octal digit by its 3-bit binary representation. (If an octal digit is less than 4, its 3-bit binary representation will begin with one or more 'leading' zeros — don't leave them out!)

EXAMPLE 2.4.4

| Convert 514.7_8 to binary.

SOLUTION

$$\underbrace{5}_{101} \ \underbrace{1}_{001} \ \underbrace{4}_{100} \cdot \underbrace{7}_{111}$$

$$514.7_8 = 101001100.111_2$$

Conversion between binary and hexadecimal is similar, except that each hexadecimal digit corresponds to *four* bits.

EXAMPLE 2.4.5

| Convert 10111101001.110001_2 to hexadecimal.

SOLUTION

$$\underbrace{101}_{5} \ \underbrace{1110}_{E} \ \underbrace{1001}_{9} \cdot \underbrace{1100}_{C} \ \underbrace{0100}_{4}$$

$$10111101001.110001_2 = 5E9.C4_{16}$$

EXAMPLE 2.4.6

| Convert $B2.5D6_{16}$ to binary.

SOLUTION

$$
\begin{array}{ccccc}
\underbrace{B}_{1011} & \underbrace{2}_{0010} & \cdot & \underbrace{5}_{0101} & \underbrace{D}_{1101} & \underbrace{6}_{0110}
\end{array}
$$

$$B2.5D6_{16} = 10110010.01011101011_2$$

The hexadecimal system is commonly used in computing to represent the contents of part of the memory or a binary file in human-readable form, since each byte (consisting of 8 bits) can be represented by 2 hexadecimal digits.

2.5 *Arithmetic in non-decimal bases*

The rules for adding, subtracting, multiplying and dividing numbers by hand in bases other than 10 are the same as the rules you learnt in primary school for the decimal system; it is only the tables that are different. In this section, we look at how binary arithmetic with natural numbers can be performed by hand. In Chapter 3, we will see how computers perform arithmetic in the binary system.

The addition table in the binary system is shown in Table 2.3.

+	0	1
0	0	1
1	1	10

Table 2.3

Numbers are added in the usual column-by-column fashion. Sometimes there will be a 1 to 'carry' to the next column.

For example:

$$
\begin{array}{r}
11011_2 \\
1110_2 \\
\hline
101001_2
\end{array}
$$

In this example, the second column from the right gives $1 + 1 = 10$, so there is a 1 to carry to the third column. Similarly, there is a carry from the third to the fourth column, from the fourth to the fifth column, and from the fifth to the sixth column.

This is essentially the method used by a computer to add natural numbers.

Subtraction can also be done by the usual method, as shown in the example below. (This is not the way most computers perform subtraction. We will see in Chapter 3 how subtraction is done on a computer.)

$$
\begin{array}{r}
11011_2 \\
- \quad 1110_2 \\
\hline
1101_2
\end{array}
$$

The multiplication table in the binary system is shown in Table 2.4.

×	0	1
0	0	0
1	0	1

Table 2.4

A 'long multiplication' in binary is carried out without really doing any multiplications at all. (All you ever need to multiply by is 0 or 1.) Here is an example; notice that the row corresponding to multiplying by the fours bit of the multiplier is omitted because that bit is zero:

$$
\begin{array}{r}
11010_2 \\
\times\ \ 1011_2 \\
\hline
11010_2 \\
11010_2 \\
11010_2 \\
\hline
100011110_2 \\
\end{array}
$$

Long division is also straightforward, with the proviso that the fractional part of the result may be non-terminating, in which case you need to decide how many digits are required in the answer. At each step of the division, the divisor 'goes into' the number either once (if it is less than or equal to the number), or not at all (if it is greater than the number). For example:

$$
\begin{array}{r}
101_2 \overline{)\ 11101_2} \quad 1 \\
101_2 \\
\hline
100_2 \quad 0 \\
0_2 \\
\hline
1001_2 \quad 1 \\
101_2 \\
\hline
100_2 \\
\end{array}
$$

In this example, the integer quotient 101_2 has been calculated, leaving a remainder of 100_2.

Some calculators have the facility for converting between the bases 10, 2, 8 and 16, and for performing arithmetic in those bases. Usually only unsigned (positive) integers can be used in these calculations. Computer algebra software such as DERIVE® provides more flexible facilities for working with numbers in different bases.

Exercises

1. Write the decimal number 394.27_{10} in expanded form.

2. Convert the following binary numbers to decimal by first writing them in expanded form:

   (a) 1100101_2
   (b) 1010111.1011_2

3. Convert the following numbers from decimal to binary:

   (a) 826_{10}
   (b) 0.34375_{10}
   (c) 1604.1875_{10}
   (d) -471.25_{10}

4. Convert the following numbers from decimal to binary, with 5 digits after the point:

   (a) 0.2_{10}
   (b) 13.47_{10}

5. What is the effect on the value of a natural number if:

   (a) 0 is appended to its binary representation?
   (b) 1 is appended to its binary representation?

6. Convert the following octal and hexadecimal numbers to decimal:

   (a) 4715_8
   (b) 603.25_8
   (c) $C6E_{16}$
   (d) $2FA.8_{16}$

7. An efficient computational method for converting a natural number from a non-decimal base to decimal, known as Horner's method, is illustrated in the following example:

$$6253_8 = ((6 \times 8 + 2) \times 8 + 5) \times 8 + 3 = 3243_{10}$$

   In words: multiply the first digit by the base, add the second digit, multiply by the base, add the third digit, multiply by the base, and so on. The final step is to add the last digit.

   Use Horner's method to convert the following numbers to decimal:

   (a) 7216_8
   (b) 543517_8
   (c) $8CB2_{16}$
   (d) $E490DF_{16}$

8. Write Horner's method as an algorithm in pseudocode. Assume that the base of the number to be converted is input first, followed by the number itself, regarded as a list of digits in that base.

9. Convert the following decimal numbers to octal:

   (a) 3842_{10}
   (b) 291.9375_{10}

10. Convert the following decimal numbers to hexadecimal:

    (a) 29803_{10}
    (b) 6962.578125_{10}

11. Convert the following binary numbers to octal and hexadecimal:

    (a) 1110100110_2
    (b) 11000101.00111_2

12. Convert the following octal and hexadecimal numbers to binary:

    (a) 247_8
    (b) 31.63_8
    (c) $93B_{16}$
    (d) $AD.1C_{16}$

13. The contents of 1 byte (consisting of 8 bits) in a memory register of a computer can be represented by 2 hexadecimal digits. A *left shift* is an operation in which the 8 bits are moved one position to the left, the leftmost bit is lost, and a 0 is inserted in the rightmost position. A *right shift* is defined in a similar manner.

   For each of the following bytes, find the result (in hexadecimal) of a left shift and a right shift:

   (a) 3A   **(b)** E7

14. Perform the following calculations in binary arithmetic:

   (a) $1101101_2 + 1011110_2$
   (b) $1001101_2 + 101011_2$
   (c) $1110011_2 - 101101_2$
   (d) $1100010_2 - 1010111_2$
   (e) $10011_2 \times 1101_2$
   (f) $11010_2 \times 10101_2$
   (g) $110110_2 \div 1001_2$
   (h) $10110_2 \div 11_2$ (3 digits after the point)

15. Fill in the details in the following outline of a proof that $\sqrt{2}$ is irrational:

   1. Assume $m/n = \sqrt{2}$ where m and n are natural numbers. Explain why we can assume that m and n are not both even.
   2. Deduce that $m^2 = 2n^2$, and hence explain why m must be even.
   3. Let $m = 2k$ (where k is a natural number), and deduce that n is even.
   4. Explain how this proves that $\sqrt{2}$ is irrational.

3. COMPUTER REPRESENTATION AND ARITHMETIC

3.1 Representing numbers in a computer

In Chapter 2, we looked at how numbers can be represented in the binary number system. In this chapter, we shall use the techniques we developed in Chapter 2 to investigate the ways in which numbers are represented and manipulated in binary form in a computer.

Numbers are usually represented in a digital computer as sequences of bits of a fixed length. Integers and real numbers are handled in different ways, and so we need to deal with the two cases separately. The details vary from one type of computer to another, but our aim here is to gain an overview of the way in which computers handle numbers, rather than to study any particular type of machine in detail.

3.2 Representing integers

In most computers, an integer (positive, negative or zero) is stored in a sequence of bytes (usually 2 or 4), where each byte consists of 8 bits. If 2 bytes are used, for example, the registers in which an integer is stored can be visualised in the following way, where each box represents the storage location for one bit:

Since there are two possible values (0 or 1) for each bit, the number of different integers that can be stored in 16 bits is:

$$2 \times 2 \times 2 \times \ldots \times 2 \ (16 \text{ times})$$
$$= 2^{16}$$
$$= 65536$$

For example, it is possible to store any integer n that falls in the range $-32768 \le n \le 32767$, by assigning a unique 16-bit pattern to each integer in that range. (This assumes that no bit patterns have been set aside for special purposes, such as error codes to signal when overflow has occurred during a calculation.) Other ranges of values containing 65536 integers could be used instead, such as $0 \le n \le 65535$, but a range that includes approximately the same number of positive and negative integers is generally the most useful.

The most commonly used way of representing an integer n using 16 bits is as follows:

1. The first bit is the *sign bit*; it is 0 if n is zero or positive, and 1 if n is negative.
2. If $n \ge 0$, the remaining 15 bits are the binary representation of n as described in Chapter 2 (with leading zeros if necessary, to bring the number of bits up to 15). If $n < 0$, the remaining bits are the binary representation of the non-negative integer $n + 32768$.

Adding 32768 to n when n is negative might seem to be an unnecessary complication, and you might wonder why the remaining bits could not instead be set equal to the 15-bit binary representation of $|n|$.[1] The reason is that by adding 32768 we obtain a representation of n that allows a simpler process to be used to carry out arithmetic with integers, as we will see in the next section.

The addition of 32768 is actually easier to perform *after* the negative integer n has been converted to binary, as the following example illustrates. Suppose we want to find the 16-bit computer representation of -6772. Converting -6772 to binary (using 15 bits, including leading zeros), we obtain $-6772_{10} = -001101001110100_2$. Now, since the binary representation of 32768 is 1000000000000000_2, we need to perform the following calculation:

$$
\begin{array}{r}
1000000000000000_2 \\
-\quad 001101001110100_2 \\
\hline
110010110001100_2
\end{array}
$$

There is a short-cut method of doing this subtraction:

1. All the zeros at the right-hand end of the number being subtracted remain as zeros in the answer.
2. The rightmost 1 of the number remains as a 1 in the answer.
3. All the other bits change (0 changes to 1, 1 changes to 0).

The result of applying these steps is called the *2's complement* of the original number with respect to the base 2. The '2's' is included in the name to

---

[1] $|n|$ denotes the *absolute value* of n, defined by:

$$
|n| = \begin{cases} n & \text{if } n \ge 0 \\ -n & \text{if } n < 0 \end{cases}
$$

For example, $|3| = 3$ and $|-3| = 3$.

distinguish this type of complement from the 1's complement (still with respect to base 2). The 1's complement of a binary number is one less than its 2's complement, and is obtained simply by changing each bit of the number (0 changes to 1, 1 changes to 0). We will not use 1's complements in what follows.

More generally, let n be an integer in the range $1 \leq n \leq 2^k - 1$, written in its k-bit binary representation. The k-bit[2] 2's complement of n is $2^k - n$ written in its k-bit binary representation. It is obtained by applying the above steps to n.

The short-cut method cannot be used to find the computer representation of -32768, because the binary representation of 32768 has 16 bits, not 15. In this case the result of the subtraction is a string of 15 zeros.

EXAMPLE 3.2.1

Find the 16-bit computer representations of the following integers:

(a) 9843   (b) −15728   (c) −4961

SOLUTION

(a) The sign bit is 0, and $9843_{10} = 010011001110011_2$ (using 15 bits). The 16-bit representation is:

$$00100110\ 01110011$$

(b) The sign bit is 1.
    Convert 15728 to binary: $15728_{10} = 011110101110000_2$
    Find the 2's complement: 100001010010000_2
    The 16-bit representation is:

$$11000010\ 10010000$$

(c) The sign bit is 1.
    Convert 4961 to binary: $4961_{10} = 001001101100001_2$
    Find the 2's complement: 110110010011111_2
    The 16-bit representation is:

$$11101100\ 10011111$$

If the computer uses 4 bytes (32 bits) to store integers instead of 2, any integer n in the range $-2147483648 \leq n \leq 2147483647$ can be represented, because $2147483648 = 2^{31}$. The method for finding the representation is essentially unchanged, except that 2147483648 takes the place of 32768, and the binary conversion must be taken to 31 bits instead of 15.

---

[2]The number of bits is usually not stated explicitly, but is implied by the number of bits given in n.

3.3 *Arithmetic with integers*

In order for us to investigate how a computer adds and subtracts integers, it is convenient to work with examples in which integers are represented using a small number of bits — far smaller than any computer would use in practice. The examples in this section will all involve an imaginary computer in which integers are stored as four bits. Only the integers from –8 to 7 can be represented on this computer. Table 3.1 gives a complete list of these integers, together with their representations according to the rules given in the previous section.

Integer	Representation
–8	1000
–7	1001
–6	1010
–5	1011
–4	1100
–3	1101
–2	1110
–1	1111
0	0000
1	0001
2	0010
3	0011
4	0100
5	0101
6	0110
7	0111

Table 3.1

(Note that to calculate the representation of –8, we cannot use the short-cut rule for finding the 2's complement.)

By examining Table 3.1, we can make the following observations:

- For the non-negative integers (0 to 7), the last 3 bits of the computer representation form the 3-bit binary representation of the integer. Since the sign bit is 0 for these integers, the computer representation is just the 4-bit binary representation of the integer.
- For a negative integer n (–8 to –1), the last 3 bits of the computer representation form the 3-bit binary representation of $n + 8$. Since the sign bit is 1, with a place value of 8 if the computer representation is treated as a binary number, the computer representation is the 4-bit representation of $n + 16$. (For example, the computer representation of –3 is 1101, which is the binary representation of 13, and $13 = (-3) + 16$.)

Suppose now that we want to add two integers on this computer. Notice first that this will not always be possible; we cannot add 6 and 7, for example, since the result is too large to be represented. Provided we restrict ourselves to numbers that *can* be added, however, addition can be done simply by adding the computer representations in the usual way that binary numbers are added, except that *if a 1 arises in the fifth column from the right, then it is ignored.*

EXAMPLE 3.3.1

Verify that the following additions are done correctly on the 4-bit computer:

(a) $2 + 3$ (b) $(-4) + 7$
(c) $(-3) + (-4)$ (d) $(-6) + 5$

SOLUTION

We write the computation in two columns, with the left column containing the numbers to be added and their sum, and the right column containing the corresponding computer representations.

(a)
$$
\begin{array}{rl}
2 & 0010 \\
3 & 0011 \\
\hline
5 & 0101
\end{array}
$$

(b)
$$
\begin{array}{rl}
-4 & 1100 \\
7 & 0111 \\
\hline
3 & 1\ 0011
\end{array}
$$

Ignoring the 1 in the leftmost column of the result gives 0011, which is the computer representation of 3, the correct answer to $(-4) + 7$.

(c)
$$
\begin{array}{rl}
-3 & 1101 \\
-4 & 1100 \\
\hline
-7 & 1\ 1001
\end{array}
$$

Ignoring the 1 gives 1001, the computer representation of −7.

(d)
$$
\begin{array}{rl}
-6 & 1010 \\
5 & 0101 \\
\hline
-1 & 1111
\end{array}
$$

It will be easier to see why this process works after we have studied modular arithmetic in Chapter 12, but the general idea can be explained as follows. The value of the representation of an integer on this computer is always the integer itself or 16 more than the integer. Therefore, when two representations are added, the value of the result either equals the answer to the addition, or exceeds it by 16 or 32 (= 2 × 16). Ignoring a 1 in the fifth column from the right is equivalent to subtracting 16. Therefore, when the process is complete, if the result is not the correct answer to the addition, it must differ from the correct answer by a multiple of 16. As long as the sum of the two numbers falls in the allowed range from –8 to 7, the process must give the correct result, because the difference between any two integers in the allowed range is less than 16.

If we try to use this method to add two integers that give a result outside the allowed range, we will obtain the wrong answer. For example, you can check that using the method to add 6 and 7 gives an answer of –3. This *overflow* problem can be easily detected, however. If adding two positive integers appears to give a negative result, or if adding two negative integers appears to give zero or a positive result, then overflow must have occurred. Some (but not all) applications software that uses integer arithmetic, including some compilers for programming in high-level languages, will generate an appropriate error message whenever this happens. This is not the case if you are programming in a low-level (assembly) language, so you need to be aware of the possibility of overflow and know how to deal with it.

Subtraction of one integer from another using the computer representation is quite straightforward, once we have made two observations:

- A number can be subtracted by adding its negative, that is, $a - b = a + (-b)$.
- The representation of the negative of an integer is the 2's complement of the representation of the integer.

Subtraction of a number can therefore be carried out by *adding the 2's complement of the number*. This means that any subtraction problem is easily converted into an addition problem. The advantage of doing subtraction in this way is that a computer does not need 'subtracters' as well as 'adders' in its circuitry (although it does need circuitry to calculate 2's complements).

The need to be alert to the possibility of overflow applies to subtraction as well as to addition.

EXAMPLE 3.3.2

| Evaluate $5 - 3$ on the 4-bit computer.

SOLUTION

The representations of 5 and 3 are 0101 and 0011 respectively. The 2's complement of 0011 is 1101. Now carry out the addition:

$$
\begin{array}{r}
0101 \\
1101 \\
\hline
1\ 0010
\end{array}
$$

Ignore the leftmost 1 to obtain 0010, which is the representation of 2.

Subtraction using complements can be performed in any base. The subtraction of numbers written in decimal notation using this method is explored in Exercises 3 and 4 at the end of the chapter.

3.4 *Representing real numbers*

The representation of real numbers is more complicated than that of integers, not just in a computer but in printed form as well. In Chapter 2, we dealt with real numbers expressed in the decimal system with digits before and after the decimal point. Not only is this not a convenient representation for a computer, it is often not suitable as a written representation either. One reason for this is that in many problems of practical interest, it is necessary to deal with real numbers that are extremely small or extremely large, and that occur as a result of performing a physical measurement to a certain degree of accuracy. This is a subtle point, but an important one.

An example will help to make this clear. The mass of the Earth is sometimes quoted as 0.5976×10^{25} kg. Notice that this number is not written as 5976 followed by 21 zeros. Not only would it be inconvenient to write the number in that way, it would also not be correct to do so. The figure of 0.5976×10^{25} kg would have been calculated from physical measurements, and no measurement is ever free of error. In this case, the accuracy of the figure is implied by the way it is written. We cannot say that the mass of the Earth is exactly 0.5976×10^{25} kg, only that it lies somewhere between 0.59755×10^{25} kg and 0.59765×10^{25} kg. It might turn out when more accurate measurements become available that a closer value is, say, 0.59758×10^{25} kg.

The notation used in this example is called *exponential* notation (or 'powers-of-ten' notation). The use of exponential notation is especially appropriate in scientific work, where the numbers arise from physical measurements.

There is some terminology associated with exponential notation. In the number 0.5976×10^{25}, 0.5976 is the *mantissa*, 10 is the *base* (or radix), and 25 is the *exponent*. If the mantissa m lies in the range $0.1 \leq m < 1$, as it does in this example, the representation is said to be *normalised*. (If we had written 0.005976×10^{27}, for example, the number would still be in exponential notation, but it would not be normalised.)

Exponential notation is similar to the way in which real numbers are represented in a computer, but there are some differences. The main difference, not surprisingly in view of the way in which computers manipulate data, is that a computer uses powers of 2 rather than 10. It is also usual for the exponent to be stored in a modified form, as we will see shortly.

Before the computer representation of a real number can be found, the number must first be converted to binary form and then be expressed in *normalised binary exponential form*. A real number is expressed in normalised binary exponential form if it is expressed in the form:

$$\pm m \times 2^{e}$$

where the mantissa m is written in its binary representation and lies in the range $0.1_2 \leq m < 1_2$, and the exponent e is an integer written in its decimal representation. (The base 2 is also in decimal notation, of course.)

Note that zero cannot be expressed in normalised binary exponential form, because $m = 0$ would fall outside the allowed range of values of m.

EXAMPLE 3.4.1

Express the following numbers in normalised binary exponential form:

(a) 11001.101_2 (b) 0.000110111_2

SOLUTION

(a) Move the point five places to the left to obtain the mantissa. The exponent is 5.

$$11001.101_2 = 0.11001101_2 \times 2^5$$

(b) Move the point three places to the right to obtain the mantissa. The exponent is −3.

$$0.000110111_2 = 0.110111_2 \times 2^{-3}$$

A real number is typically stored in a computer as 4 bytes (32 bits) or 8 bytes (64 bits)[3]. The first bit is the sign bit, and the remaining bits are divided between the exponent and the mantissa. The number of bits used for the exponent determines the *range* of numbers that can be represented, while the number of bits used for the mantissa determines the *precision* with which the numbers are represented. If the total number of bits is fixed, there is a trade-off between range and precision. A common format in modern computers is 8 bits for the exponent and 23 bits for the mantissa.

It is usually not the binary representation of the exponent itself that is stored, but a number called the *characteristic*. The characteristic is a non-negative integer obtained by adding to the exponent a number called the *exponent bias*. The exponent bias is typically $2^{n-1} - 1$, where n is the number of bits available to store the characteristic. The reason for storing the exponent in this way is that it allows the computer to use simpler algorithms to perform arithmetic with real numbers, in much the same way that integer arithmetic is simplified by representing negative integers in 2's complement form.

EXAMPLE 3.4.2

Find the 32-bit computer representations of

(a) $0.11001101_2 \times 2^5$ (b) $0.110111_2 \times 2^{-3}$

where 8 bits are used for the characteristic, and the exponent bias is $2^7 - 1$.

---

[3]If the two different formats are available on the same machine, they are referred to as *single precision* and *double precision* respectively.

SOLUTION

(a) The sign bit is 0. The characteristic is the 8-bit binary representation of $5 + 2^7 - 1$, which can be obtained by adding 10000000_2 (the binary representation of 2^7) to the binary representation of 4, giving 10000100. The mantissa is extended to 23 bits by appending trailing zeros, giving 11001101000000000000000. (Note that the zero before the point is not stored.) The computer representation is:

$$01000010\ 01100110\ 10000000\ 00000000$$

(b) The sign bit is 0. The characteristic is $-3 + 2^7 - 1$, represented as an 8-bit binary number. The simplest way to calculate the characteristic here is to find the 7-bit 2's complement of the binary representation of $4\ (= 3 + 1)$, and adjoin a leading zero:

Binary representation of 4: 0000100_2
2's complement: 1111100_2
Characteristic: 01111100
The computer representation is:

$$00111110\ 01101110\ 00000000\ 00000000$$

Putting together what we have established, the process for finding the computer representation of a real number can be described as follows:

1. Convert the number to binary form, working to the precision required by the number of bits used for the mantissa.
2. Express the binary number in normalised binary exponential form.
3. Calculate the characteristic.
4. Write down the computer representation.

EXAMPLE 3.4.3

Find the 32-bit computer representation of -1873.42, where 8 bits are used for the characteristic, and the exponent bias is $2^7 - 1$.

SOLUTION

Using the methods of Chapter 2, convert -1873.42 to a binary number with 23 bits:

$$-1873.42_{10} = -11101010001.011010111000_2$$

Express the result in normalised binary exponential form:

$$-11101010001.011010111000_2 = -0.11101010001011010111000 \times 2^{11}$$

Sign bit: 1
Characteristic: 10001010
Computer representation:

11000101 01110101 00010110 10111000

Example 3.4.3 illustrates the important fact that *the computer representation of a real number is not exact*, because the conversion to binary is truncated according to the precision available. Further inaccuracy can occur as a result of round-off errors when arithmetic is performed with real numbers. This is in contrast to the situation for integers, which are stored and manipulated exactly in a computer.

One practical consequence of this fact is that it is risky to test for equality of real numbers (in an **If-then** statement, for example) in a computer program. If x and y are real numbers, it is safer to test for approximate equality by testing whether $|x-y|$ is less than some small positive number. For example, we might write:

$$\textbf{If } |x-y| < 10^{-6} \textbf{ then } \ldots$$

to test whether x and y are approximately equal.

What is the range of real numbers that can be represented? Suppose a computer stores real numbers as 32 bits, with 8 bits for the characteristic, and an exponent bias of $2^7 - 1$. The characteristic can take values from 0 to $2^8 - 1$, so the exponent must lie within the range from $0 - (2^7 - 1)$ to $2^8 - 1 - (2^7 - 1)$, that is, from -127 to 128. The mantissa ranges from 0.1_2 to (just under) 1_2. Therefore the range of positive real numbers that can be represented is from $0.1_2 \times 2^{-127}$ to $1_2 \times 2^{128}$, or about 0.29×10^{-38} to 0.34×10^{39} in decimal notation. Negative real numbers whose absolute values fall in this range can also be represented. If the result of a computation falls outside the allowed range, overflow or underflow[4] occurs. The numbers that can be represented are indicated on the number line shown in Figure 3.1; note that the line is not drawn to scale.

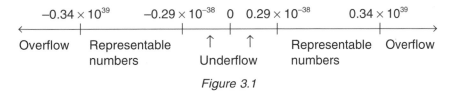

Figure 3.1

---

[4]Underflow occurs when the absolute value of the result of a computation involving real numbers is less than the smallest positive number that can be represented. Depending on the context in which the computation occurs, it may or may not be appropriate to approximate the result of an underflow by zero.

EXAMPLE 3.4.4

Find the approximate range of positive real numbers that can be represented if 10 bits are available for the characteristic, and the exponent bias is $2^9 - 1$.

SOLUTION

The characteristic can take values from 0 to $2^{10} - 1$, so the exponent must lie within the range from $0 - (2^9 - 1)$ to $(2^{10} - 1) - (2^9 - 1)$, that is, from -511 to 512. Therefore the range of positive real numbers that can be represented is from $0.1_2 \times 2^{-511}$ to $1_2 \times 2^{512}$. Some calculators give an error message if an attempt is made to evaluate 2^{-511} directly, but the calculation can be done using base 10 logarithms:

$$\log_{10} 2^{-511} = -511 \log_{10} 2$$
$$= -153.8263278$$

Therefore:

$$2^{-511} = 10^{-153.8263278}$$
$$= 10^{-153} \times 10^{-0.8263278}$$
$$= 0.149 \times 10^{-153}$$

We can calculate 2^{512} in a similar fashion. The final result is that numbers from 0.75×10^{-154} to 0.13×10^{155} can be represented on this machine.

The number zero is not included among the real numbers whose representations we have been discussing, because it cannot be expressed in normalised form. The representation of zero has to be treated as a special case, using a string of bits that cannot be interpreted as any other number.

Some variations on this representation are often used in practice. On some machines, the range of real numbers that can be represented may be a little smaller than we have described, because the exponents at the extremes of the range have been set aside to denote exceptional conditions such as overflow. Sometimes the leading bit of the mantissa is omitted from the representation, because it is always 1 and therefore carries no information; this yields a greater degree of precision by making one extra bit available for the mantissa.

3.5 *Arithmetic with real numbers*

We will not look at how a computer does arithmetic with real numbers using their computer representations, because performing these calculations by hand is tedious. However, a general idea of what the process involves can be gained by looking at how arithmetic is done with decimal numbers in normalised exponential form.

The rules for arithmetic in normalised exponential form are as follows:

To add or subtract:

1. Write the numbers in (non-normalised) exponential form with the same exponent, using mantissas less than 1.
2. Add or subtract the mantissas to obtain the mantissa of the answer. The common exponent is the exponent of the answer.
3. Normalise the answer if necessary.

To multiply or divide:

1. Multiply or divide the mantissas to obtain the mantissa of the answer.
2. Add or subtract the exponents to obtain the exponent of the answer.
3. Normalise the answer if necessary.

In order to imitate as closely as possible the way in which a computer would carry out the computations, we will assume that the number of decimal digits available for the mantissa in the normalised form is fixed, and that the answer is rounded off to that number of digits.

EXAMPLE 3.5.1

Perform the following computations with the aid of a calculator, assuming a precision of 4 decimal places in the mantissa:

(a) $0.4932 \times 10^3 + 0.2881 \times 10^4 - 0.3096 \times 10^4$

(b) $(0.2174 \times 10^{-5}) \times (0.1482 \times 10^7) \div (0.9497 \times 10^4)$

SOLUTION

(a) $0.4932 \times 10^3 + 0.2881 \times 10^4 - 0.3096 \times 10^4$
$$= 0.0493 \times 10^4 + 0.2881 \times 10^4 - 0.3096 \times 10^4$$
$$= 0.0278 \times 10^4$$
$$= 0.2780 \times 10^3$$

(b) $(0.2174 \times 10^{-5}) \times (0.1482 \times 10^7) \div (0.9497 \times 10^4)$
$$= 0.03393 \times 10^{-2}$$
$$= 0.3393 \times 10^{-3}$$

Example 3.5.1 illustrates how round-off errors can arise when real number calculations are performed.

3.6 *Binary coded decimal representation*

Modern digital computers work in the binary number system. However, some early computers performed calculations in the decimal system, using a binary code for each decimal digit. This system, known as *binary coded decimal*, or BCD, is also the system commonly used in electronic calculators.

The BCD code of each decimal digit from 0 to 9 is its 4-bit binary representation, as shown in Table 3.2.

Digit	BCD code
0	0000
1	0001
2	0010
3	0011
4	0100
5	0101
6	0110
7	0111
8	1000
9	1001

Table 3.2

The BCD representation of a non-negative integer is obtained by replacing each decimal digit by its BCD code.

For example, the BCD representation of 8365 is:

$$1000 \ 0011 \ 0110 \ 0101$$

BCD arithmetic is essentially decimal arithmetic done according to the familiar rules, but with BCD codes in place of the digits. We shall illustrate this by investigating the addition of two numbers in the BCD system.

When we add two numbers in the decimal system, we add pairs of digits a column at a time, and sometimes there is a 'carry' to the next column to the left. Essentially the same process is used in the BCD system.

How are BCD digit codes added? By examining Table 3.2, we see that any two BCD codes can be added using the rules of binary addition to give the BCD code of the result, provided that the result is less than 10 (e.g. adding the codes for 3 and 4 gives the code for 7: $0011_2 + 0100_2 = 0111_2$). If the result is 10 or more (1010_2 or more in binary), there is a (decimal) digit to 'put down' and a 1 to 'carry'. The digit to be 'put down' can be found using the rule: *add six (0110_2) and disregard the leftmost 1*. (The leftmost 1 can be thought of as the '1 to carry'.)

For example, adding 0101_2 and 0111_2 (5 and 7) as binary numbers gives 1100_2. Now add 0110_2 to obtain 10010_2. Interpret this binary number as

0001 0010, that is, put down 0010 (the BCD representation of 2) and carry 0001. On replacing each BCD code with its corresponding decimal digit, we obtain 12, the correct answer to the addition.

EXAMPLE 3.6.1

| Calculate 274 + 163 in BCD.

SOLUTION

The decimal and BCD calculations are shown side by side to show the relationship between them.

$$
\begin{array}{rl}
274 & \text{0010 0111 0100} \\
163 & \text{0001 0110 0011} \\
\hline
     & \text{0011 1101 0111} \\
     & \text{\qquad\quad 0110} \\
\hline
     & \text{1 0011} \\
\hline
437 & \text{0100 0011 0111}
\end{array}
$$

BCD calculations take longer to perform than the corresponding calculations in the binary system. The advantage of the BCD representation is that it avoids the need to convert the input from decimal to binary and the output from binary to decimal.

Exercises

1. Find the 2's complements of the following 8-bit binary numbers:
   (a) 11010100_2 (b) 01101001_2

2. Write an algorithm in pseudocode for finding the 2's complement of a positive binary integer. (Assume that the integer is input as a sequence of bits: $b_1 b_2 b_3 \ldots b_n$.)

3. The 10's complement of a positive decimal integer n is $10^k - n$, where k is the number of digits in the decimal representation of n. It can be calculated in the following way:

   1. All the zeros at the right-hand end of the number remain as zeros in the answer.
   2. The rightmost non-zero digit d of the number is replaced by $10 - d$ in the answer.
   3. Each other digit d is replaced by $9 - d$.

Find the 10's complements of the following decimal numbers using the rules given above, and check your answers by evaluating $10^k - n$ on a calculator:

(a) 3296   (b) 10350

4. Subtraction $a - b$ (with $a > b$) can be performed in the decimal system by adding the 10's complement of b to a and ignoring the leftmost 1 of the answer.

(a) Evaluate $39842 - 17674$ using this method.
(b) Explain why the method works.

5. Find the 16-bit computer representations of the following integers:

(a) 29803   (b) –8155

6. The maximum unsigned integer on a CRAY-1 computer is approximately 2.8×10^{14}. How many bits are used to store unsigned integers on a CRAY-1? (Unsigned integers are non-negative integers stored without a sign bit.)

7. Verify that $3 + (-5)$ is evaluated correctly on the 4-bit computer.

8. Evaluate $7 - 6$ on the 4-bit computer.

9. Express 1101110100.1001_2 in normalised binary exponential form. Hence find its 32-bit computer representation, assuming 8 bits are used for the characteristic, and the exponent bias is $2^7 - 1$.

10. Find the 32-bit computer representations of the following numbers, assuming 8 bits are used for the characteristic, and the exponent bias is $2^7 - 1$:

(a) 5894.376   (b) –0.0387

11. Repeat Question 9 for a 32-bit computer in which 12 bits are used for the characteristic, and the exponent bias is $2^{11} - 1$.

12. Find, in decimal form, the approximate range of positive real numbers that can be represented in 64 bits, where 11 bits are used for the characteristic, and the exponent bias is $2^{10} - 1$.

13. The algorithm below represents an attempt to print out a table of cubes of numbers from 0.1 to 10 in steps of 0.1. Explain the problem that might arise if the algorithm is implemented on a computer:

1. $x \leftarrow 0.0$
2. **Repeat**
    2.1. $x \leftarrow x + 0.1$
    2.2. $x\_cubed \leftarrow x^3$
    2.3. Output x, x_cubed
    until $x = 10.0$

Rewrite the algorithm so that the problem is avoided.

14. Perform the following decimal computations, assuming a precision of 4 decimal places in the mantissa:

    (a) $0.8463 \times 10^6 + 0.7012 \times 10^8$
    (b) $(0.3315 \times 10^{-5}) \times (0.2089 \times 10^9)$
    (c) $(0.5160 \times 10^3) \div (0.1329 \times 10^4) - (0.3816 \times 10^0)$

15. The following statement is sometimes made about real number arithmetic: 'Precision is lost when two almost equal numbers are subtracted'. Explain this statement with reference to calculations with real numbers in exponential form.

16. Perform the following calculations in BCD arithmetic:

    (a) $3711 + 5342$ (b) $2859 + 3264$

4. LOGIC

4.1 *Logic and computing*

In this chapter, we will introduce the study of logic from a mathematical point of view. Mathematical logic finds applications in many areas of computing. The laws of logic are employed in the design of the digital circuitry in a computer. Logical expressions occur as conditions in the control structures in algorithms and computer programs, and in the commands used for querying databases. Expert systems employing knowledge-based software use rules of logical inference to draw conclusions from known facts. Formal specification documents, which state in a precise way what computer systems are required to do, are written in specification languages, such as Z, which use the theory and notation of symbolic logic.

We will begin this chapter by looking at examples involving everyday English sentences. This will be followed by an introduction to the more formal mathematical approach used in propositional and predicate logic.

4.2 *Propositions*

The fundamental objects we work with in arithmetic are numbers. In a similar way, the fundamental objects in logic are *propositions*.

Definition

A *proposition* is a statement that is either true or false. Whichever of these (true or false) is the case is called the *truth value* of the proposition.

Here are some examples of English sentences that are propositions:

'Canberra is the capital of Australia.'
'There are 8 days in a week.'
'Isaac Newton was born in 1642.'
'5 is greater than 7.'
'Every even number greater than 2 can be expressed as the sum of two prime numbers.'

The first and third of these propositions are true, and the second and fourth are false. It is not known at present whether the fifth proposition is true or false.[1]

The following sentences are not propositions:

'Where are you going?'
'Come here.'
'This sentence is false.'

The first sentence is a question and the second is a command, so clearly neither is a proposition.

The third sentence is rather more subtle. It is a *self-referential* statement (i.e. it makes a statement about itself). Although at first sight it appears to be a proposition, we run into difficulty when we try to determine whether it is true or false. If we assume it is true, we find the sentence telling us that it is false, which contradicts our assumption. But assuming the sentence is false doesn't work either, because if what the sentence is telling us is false, then the sentence is true! The sentence is an example of a *paradox*, and the only way to avoid the difficulty is simply not to admit the sentence as a proposition. In fact, we will not allow self-referential statements at all in our work in logic. (This does not mean that there is no place for self-reference in logic; in fact, some of the most important results in modern logic involve self-referential propositions.)

Now, what about sentences like these?

'Anne is tall.'
'Ice cream is delicious.'
'$x > 5$.'

In some textbooks, sentences like the first two are not regarded as propositions, because it could be argued that their truth values are not well defined. The first sentence refers to someone called Anne (Anne who?), and states that she is tall (just how tall is 'tall'?). The second sentence is clearly a matter of personal opinion. Quite frankly, it is not worthwhile arguing about whether these sentences are propositions or not. We will feel free to use sentences like these in some of the examples that follow. In practice, the kinds of propositions that arise when logic is applied to mathematics and computing are always precisely defined anyway, so no difficulties should arise.

The last of the three sentences given above is an example of a *predicate*. A predicate is a statement containing one or more variables; it cannot be assigned a truth value until the values of the variables are specified. We will investigate predicate logic in the final section of this chapter.

Statements containing variables commonly occur in algorithms and computer programs. For example, an algorithm might contain the statement '$x > 5$' as the condition in a control structure such as an **If-then**. In this case, however, the truth value of the statement is determined when the line is executed while

[1]The fifth proposition is known as Goldbach's conjecture, after the German mathematician Christian Goldbach (1690–1764).

the program is being run with a particular set of inputs, so statements of this type can be treated as propositions.

4.3 *Connectives and truth tables*

Logic is not concerned with determining the truth values of propositions of the kind we have seen so far. (The truth value of 'Canberra is the capital of Australia' is a question of geography, not logic.) The next example is different, however:

> 'If Brian and Angela are not both happy, then either Brian is not happy, or Angela is not happy.'

We do not need to know whether Brian is happy, or whether Angela is happy, in order to determine whether the proposition is true; with a moment's thought, we can see it *must* be true *because of its logical structure*. In fact, any sentence with the same logical structure must be true; for example:

> 'If 2 and $\sqrt{2}$ are not both rational numbers, then either 2 is not a rational number or $\sqrt{2}$ is not a rational number.'

It is the structure of propositions such as these that we study in propositional logic.

The sentence about Brian and Angela is an example of a *compound proposition*. It is built up from the *atomic* propositions 'Brian is happy' and 'Angela is happy' using the words **and**, **or**, **not** and **if-then**. These words are known as *connectives*. As we will see, the role of connectives in logic is analogous to the role played by operations such as + and × in algebra.

The study of the structure of compound propositions is made easier by the use of symbols for atomic propositions and connectives. We will use lower-case letters such as p, q and r to denote atomic propositions. There are five connectives that we will use in our work; they are listed in Table 4.1, together with their symbols.

Connective	Symbol
and	$\wedge$
or	$\vee$
not	$\neg$
if-then	$\rightarrow$
if-and-only-if	$\leftrightarrow$

Table 4.1

The connectives **if-then** and **if-and-only-if** are also known as **implies** and **is-equivalent-to** respectively.

With the exception of **not**, the symbols for these connectives are written between the two operands (the propositions they connect); for example, if p denotes the proposition 'Today is Monday', and q denotes the proposition 'It is raining', then we can write $p \wedge q$ to denote the proposition 'Today is Monday and it is raining'. The symbol ¬ **(not)** is placed before the proposition to which it applies; thus, ¬p means 'Today is not Monday'.

The connective **and** can be formally defined by stating the truth value of the proposition $p \wedge q$ for each possible combination of the truth values of the propositions p and q. The other connectives can be defined in a similar manner. This information is usually presented in the form of a *truth table*. The truth table for **and** is shown in Table 4.2.

p	q	$p \wedge q$
T	T	T
T	F	F
F	T	F
F	F	F

Table 4.2

The truth values 'true' and 'false' are denoted in the table by T and F respectively. The first two columns of the table contain all four possible combinations of the truth values of the two propositions p and q. The truth table reflects our everyday understanding of what **and** means — if p and q are both true then $p \wedge q$ is true, otherwise $p \wedge q$ is false.

The word **or** is used in English in two different ways. If you are offered tea or coffee, you are expected to choose one or the other, but not both! On the other hand, if a discount is available to anyone who is a student or a pensioner, it is presumably available to someone who is both a student and a pensioner. In the first example, **or** is used *exclusively*, while in the second example it is used *inclusively*. By convention, **or** in logic (and in computing and mathematics generally) means 'inclusive-**or**' unless the contrary is stated explicitly. The symbol ∨ always means 'inclusive-**or**'; thus $p \vee q$ means 'p or q or both'. If we wanted to use 'exclusive-**or**' (or **xor**, as it is sometimes called) in our work, we would have to define it as another connective, but we will not do this here.

The truth table for **or** is shown in Table 4.3.

p	q	$p \vee q$
T	T	T
T	F	T
F	T	T
F	F	F

Table 4.3

The truth table for **not** is straightforward; it is shown in Table 4.4.

p	$\neg p$
T	F
F	T

Table 4.4

It takes a little more thought to construct the truth table for **if-then**. Suppose your lecturer makes the following claim:

'If you pass this subject, then you will progress to the next year of your course.'

Consider the different possibilities that could occur. If you pass the subject and progress to the next year of your course, then clearly your lecturer's statement is true. If you pass the subject but don't progress to the next year of your course, then you could accuse your lecturer of making a false statement.

What happens if you fail the subject? Whether or not you progress to the next year of your course, you could not accuse your lecturer of making a false statement. (The statement only said what would happen if you passed, not what would happen if you failed.) We treat the lecturer's statement as true in these cases.

The truth table for **if-then** is shown in Table 4.5.

p	q	$p \rightarrow q$
T	T	T
T	F	F
F	T	T
F	F	T

Table 4.5

While the argument given above is intended to make the truth table for **if-then** appear reasonable, the truth table itself is really the *definition* of the connective **if-then** in logic. It follows that we can assign truth values to some rather strange English sentences. For example, the sentence 'If snow is white then lions roar' is a true proposition (according to the first line of the truth table), even though the whiteness of snow does not cause lions to roar. Another example is provided by the sentence: 'If Paris is in Germany then grass is purple'; according to the last line of the table, the sentence is true.

It is important to understand the difference between the connective **if-then** and the control structure **If-then**, which we encountered in our study of algorithms. When **If-then** is used as a control structure, the sentence following **then** is an *instruction*, not a proposition.

Finally, the connective **if-and-only-if** is true precisely when the two propositions have the same truth value (both true or both false). Its truth table is shown in Table 4.6.

p	q	$p \leftrightarrow q$
T	T	T
T	F	F
F	T	F
F	F	T

Table 4.6

For example, the proposition 'Birds have three legs if and only if $2 + 2 = 5$' is true, because the two propositions from which it is built up are both false.

4.4 *Compound propositions*

We now have the notation we need in order to be able to write compound propositions in symbolic form. Example 4.4.1 shows how this is done.

EXAMPLE **4.4.1**

Express the proposition 'Either my program runs and it contains no bugs, or my program contains bugs' in symbolic form.

SOLUTION

Let p denote the statement: 'My program runs'.
Let q denote the statement: 'My program contains bugs'.
Then the proposition can be written in symbolic form as follows:

$$(p \wedge \neg q) \vee q$$

Notice in this example how parentheses are used to group sub-expressions within the whole expression, just as in arithmetic and algebra. If the sub-expressions had been grouped differently, the meaning would have been different: $p \wedge (\neg q \vee q)$ means 'My program runs, and either it does not contain bugs or it contains bugs'. Parentheses should always be used to group sub-expressions in compound propositions, in order to avoid any ambiguity in the meaning. (There is one exception — by convention, the symbol $\neg$ immediately preceding a proposition applies only to that proposition. Thus $\neg p \wedge q$ always means $(\neg p) \wedge q$ and never $\neg(p \wedge q)$.)

In Example 4.4.1, $(p \wedge \neg q) \vee q$ is the symbolic notation for a proposition; we could find its truth value if we knew the truth values of 'My program runs' and 'My program contains bugs'. In the study of logic, it is often useful to analyse expressions such as $(p \wedge \neg q) \vee q$ in which p and q are treated as variables rather than as symbols denoting specific propositions. If we do this, then $(p \wedge \neg q) \vee q$ is no longer a proposition but a *logical expression*; it becomes a proposition only if p and q are replaced by propositions. We can think of a logical expression in the same way as we think of an expression containing a variable x in algebra — the expression can't be evaluated unless x is assigned a value, but this does not prevent us from studying the expression and investigating its properties.

We can analyse the structure of the logical expression $(p \wedge \neg q) \vee q$ in the following way. The entire expression takes the form $A \vee B$, where A is the expression $p \wedge \neg q$ and B is the variable q. The connective **or** is the *principal connective* in the original expression. In turn, $p \wedge \neg q$ takes the form $C \wedge D$, where C is the variable p, D is the expression $\neg q$, and the principal connective is **and**. Finally, the principal connective in $\neg q$ is **not**. This way of breaking down the structure of an expression is useful in constructing its truth table, as we will see shortly.

Looking ahead to Chapter 11, the structure of the expression $(p \wedge \neg q) \vee q$ can be depicted using an *expression tree*, as shown in Figure 4.1.

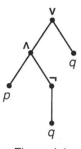

Figure 4.1

The truth value of the expression $(p \wedge \neg q) \vee q$ for each possible combination of truth values of p and q can be found by constructing a truth table.

EXAMPLE 4.4.2

Construct the truth table for the expression $(p \wedge \neg q) \vee q$.

SOLUTION

The solution is shown in Table 4.7. The first two columns of the table contain all the possible combinations of the truth values of p and q. Column 3 is obtained from Column 2 using the truth table for **not**. Column 4 is obtained from Columns 1 and 3 using the truth table for **and**. Finally,

Column 5, which contains the truth values for the entire expression, is obtained from Columns 4 and 2 using the truth table for **or**.

p	q	$\neg q$	$p \wedge \neg q$	$(p \wedge \neg q) \vee q$
T	T	F	F	T
T	F	T	T	T
F	T	F	F	T
F	F	T	F	F

Table 4.7

Notice that each column is obtained using the truth table for the principal connective in the expression at the top of the column.

If an expression contains three variables (p, q and r, say), then the table will have eight lines instead of four (there are $2^3 = 8$ different ways of allocating truth values to three expressions), but the method is the same.

Now look again at the proposition we introduced at the beginning of Section 4.3:

'If Brian and Angela are not both happy, then either Brian is not happy or Angela is not happy.'

If p and q denote respectively 'Brian is happy' and 'Angela is happy', the proposition can be expressed symbolically in the following way:

$$\neg(p \wedge q) \rightarrow (\neg p \vee \neg q)$$

The truth table for this expression is given in Table 4.8.

p	q	$p \wedge q$	$\neg(p \wedge q)$	$\neg p$	$\neg q$	$\neg p \vee \neg q$	$\neg(p \wedge q) \rightarrow (\neg p \vee \neg q)$
T	T	T	F	F	F	F	T
T	F	F	T	F	T	T	T
F	T	F	T	T	F	T	T
F	F	F	T	T	T	T	T

Table 4.8

The final column of the truth table contains only T. This tells us that the expression is always true, regardless of the truth values of p and q.

An expression that is always true, regardless of the truth values of the variables it contains, is called a *tautology*.

Now consider the following proposition:

'It is raining and windy, and it is not raining.'

Even without looking at the weather, we can tell that the proposition is false from its logical structure. We can confirm this by writing the proposition in symbolic form and constructing the truth table for the resulting expression. Using p and q to denote respectively 'It is raining' and 'It is windy', we obtain Table 4.9.

p	q	$p \wedge q$	$\neg p$	$(p \wedge q) \wedge \neg p$
T	T	T	F	F
T	F	F	F	F
F	T	F	T	F
F	F	F	T	F

Table 4.9

An expression that is always false, regardless of the truth values of the variables it contains, is called a *contradiction*.

4.5 *Logical equivalence*

Here is a rather complicated proposition:

'It is not the case that both the input file and the output file are not on the disk.'

The proposition below expresses the same idea more simply:

'Either the input file or the output file is on the disk.'

If we were to express these propositions symbolically, we would expect the resulting logical expressions to have the same truth table. Let p and q denote respectively the propositions 'The input file is on the disk' and 'The output file is on the disk'. Then we have the following result, in which for convenience we have combined the truth tables for the two expressions into a single table (Table 4.10).

p	q	$\neg p$	$\neg q$	$\neg p \wedge \neg q$	$\neg(\neg p \wedge \neg q)$	$p \vee q$
T	T	F	F	F	T	T
T	F	F	T	F	T	T
F	T	T	F	F	T	T
F	F	T	T	T	F	F

Table 4.10

The sixth and seventh columns are the truth tables for the first and second expressions respectively, and we can see that their truth values are the same.

Definition

> Two expressions (composed of the same variables) are *logically equivalent* if they have the same truth values for every combination of the truth values of the variables.

Informally, we could say that two expressions are logically equivalent if they yield the same truth table.

There is a subtle but important distinction between the connective **if-and-only-if** and the idea of logical equivalence. When we write $p \leftrightarrow q$, we are writing a *single* logical expression. Logical equivalence, on the other hand, is a relationship between *two* logical expressions. The two concepts are related in the following way: two expressions A and B are logically equivalent if and only if the expression $A \leftrightarrow B$ is a tautology.

Some important questions about logical equivalence arise when we consider expressions of the form $p \rightarrow q$. Such expressions are called *implications*. We investigate these questions now.

Definitions

> The *converse* of $p \rightarrow q$ is $q \rightarrow p$.
> The *contrapositive* of $p \rightarrow q$ is $\neg q \rightarrow \neg p$.

EXAMPLE 4.5.1

Write down English sentences for the converse and the contrapositive of:

'If 250 is divisible by 4 then 250 is an even number.'

SOLUTION

The sentence takes the form $p \rightarrow q$, where p denotes '250 is divisible by 4' and q denotes '250 is an even number'.

The converse is $q \rightarrow p$, which we can write in English as follows:

'If 250 is an even number then 250 is divisible by 4.'

> The contrapositive is $\neg q \to \neg p$, which we write as follows:
>
> 'If 250 is not an even number then 250 is not divisible by 4.'

Not only is the original proposition in the above example a true mathematical statement as it stands; it remains true if 250 is replaced by any other integer. This is also the case for the contrapositive. The converse, however, is false — 250 is an even number, but it is not divisible by 4. Example 4.5.1 suggests that $p \to q$ is not logically equivalent to its converse, but that it is logically equivalent to its contrapositive. We can confirm that this is the case by constructing a truth table (Table 4.11).

p	q	$p \to q$	$q \to p$	$\neg q$	$\neg p$	$\neg q \to \neg p$
T	T	T	T	F	F	T
T	F	F	T	T	F	F
F	T	T	F	F	T	T
F	F	T	T	T	T	T

Table 4.11

The columns for $p \to q$ and $\neg q \to \neg p$ are identical to each other, but they differ from the column for $q \to p$. Therefore an implication and its contrapositive are logically equivalent, while an implication and its converse are not.

4.6 *Laws of logic*

We began Section 4.5 with an example of a complicated proposition that we showed to be logically equivalent to a simpler one. Occasions often arise in practice where it is desirable to replace a logical expression with a simpler expression that is logically equivalent to it. For example, we have seen how logical expressions representing propositions can occur in algorithms and computer programs. By writing these expressions as simply as possible, we can make a program more efficient and reduce the chance of error.

In order to be able to simplify logical expressions effectively, we will first need to establish a list of pairs of expressions that are logically equivalent. We will use the symbol ≡ placed between two expressions to indicate that they are equivalent. A statement of the form $P \equiv Q$ where P and Q are logical expressions is called a *law* of logic. A list of the most important laws of logic is given in Table 4.12.

Law(s)		Name
$p \leftrightarrow q \equiv (p \rightarrow q) \wedge (q \rightarrow p)$		equivalence law
$p \rightarrow q \equiv \neg p \vee q$		implication law
$\neg\neg p \equiv p$		double negation law
$p \wedge p \equiv p$	$p \vee p \equiv p$	idempotent laws
$p \wedge q \equiv q \wedge p$	$p \vee q \equiv q \vee p$	commutative laws
$(p \wedge q) \wedge r \equiv p \wedge (q \wedge r)$	$(p \vee q) \vee r \equiv p \vee (q \vee r)$	associative laws
$p \wedge (q \vee r) \equiv$ $(p \wedge q) \vee (p \wedge r)$	$p \vee (q \wedge r) \equiv$ $(p \vee q) \wedge (p \vee r)$	distributive laws
$\neg(p \wedge q) \equiv \neg p \vee \neg q$	$\neg(p \vee q) \equiv \neg p \wedge \neg q$	de Morgan's laws
$p \wedge T \equiv p$	$p \vee F \equiv p$	identity laws
$p \wedge F \equiv F$	$p \vee T \equiv T$	annihilation laws
$p \wedge \neg p \equiv F$	$p \vee \neg p \equiv T$	inverse laws
$p \wedge (p \vee q) \equiv p$	$p \vee (p \wedge q) \equiv p$	absorption laws

Table 4.12

The first two laws in Table 4.12 allow the connectives **if-then** and **if-and-only-if** to be removed from any expression containing them. The remaining laws involve just the connectives **and**, **or** and **not**. With the exception of the first of these (the double negation law), the laws occur in pairs, in which the second law in the pair can be obtained from the first by interchanging $\wedge$ with $\vee$ and T with F. (Here, T means any true proposition, and F means any false proposition.) The second law in each pair is said to be the *dual* of the first, and vice versa. The double negation law is its own dual.

The list of laws in Table 4.12 is very comprehensive, and it might appear rather daunting at first. However, many of the laws are obvious after a moment's thought, such as the double negation law ('It is not the case that it is not raining' is a convoluted way of saying 'It is raining'), and the idempotent laws ('I am happy and I am happy' just means 'I am happy'). The less obvious ones can be checked using truth tables.

Notice that some of the laws take the same form as laws of ordinary algebra, with = replaced by $\equiv$, $\times$ by $\wedge$, + by $\vee$, 1 by T and 0 by F. The commutative, associative and identity laws are of this type, and so is the first of the distributive laws, because it corresponds to the familiar rule for 'multiplying out brackets': $x \times (y + z) = (x \times y) + (x \times z)$. This is not the case with the second distributive law, because $x + (y \times z) = (x + y) \times (x + z)$ is not a law of algebra. Working with the laws of logic can sometimes have the same 'feel' as doing algebra with numbers, but it is essential to make sure that each step in the solution to a problem can be justified using one of the laws of logic.

EXAMPLE 4.6.1

| Use a truth table to verify the first de Morgan's law: $\neg(p \wedge q) \equiv \neg p \vee \neg q$.

SOLUTION

Note that the law can be paraphrased as follows: 'If it is not the case that p and q are both true, then that is the same as saying that at least one of p or q is false'.

The truth table is shown in Table 4.13.

p	q	$p \wedge q$	$\neg(p \wedge q)$	$\neg p$	$\neg q$	$\neg p \vee \neg q$
T	T	T	F	F	F	F
T	F	F	T	F	T	T
F	T	F	T	T	F	T
F	F	F	T	T	T	T

Table 4.13

The columns for $\neg(p \wedge q)$ and $\neg p \vee \neg q$ are identical, and therefore the two expressions are logically equivalent.

The next example illustrates how the laws of logic can be applied to the problem of simplifying a logical expression. Starting with the given expression, a sequence of equivalent expressions is obtained by applying one of the laws at each step. You need to keep in mind that 'applying a law' often means replacing the variables in the law with logical expressions in order to put it into the required form.

EXAMPLE 4.6.2

Use the laws of logic to simplify the expression:

$$p \vee \neg(\neg p \to q)$$

SOLUTION

As this is our first example of simplifying an expression using the laws of logic, the solution is given in more detail than would normally be shown in practice, to demonstrate at each step exactly how the relevant law of logic from Table 4.12 has been applied.

$p \vee \neg(\neg p \to q) \equiv p \vee \neg(\neg\neg p \vee q)$ implication law (with $\neg p$ in place of p)

$\equiv p \vee \neg(p \vee q)$ double negation law

$\equiv p \vee (\neg p \wedge \neg q)$ second de Morgan's law

$\equiv (p \vee \neg p) \wedge (p \vee \neg q)$ second distributive law (with $\neg p$ and $\neg q$ in place of q and r respectively)

$\equiv T \wedge (p \vee \neg q)$ second inverse law

$\equiv (p \vee \neg q) \wedge T$ first commutative law (with T and $(p \vee \neg q)$ in place of p and q respectively)

$\equiv p \vee \neg q$ first identity law (with $(p \vee \neg q)$ in place of p)

There are no hard and fast rules for determining which law to apply at each step in this type of problem. If the connectives $\leftrightarrow$ or $\rightarrow$ appear in the given expression, they should be eliminated using the first two laws. After that, it can sometimes be a matter of trying a law to see if it helps to simplify the expression, and then trying another if it doesn't.

An important practical application of the laws of logic is the simplification of logical expressions in algorithms. De Morgan's laws are often particularly useful in this type of problem.

EXAMPLE 4.6.3

An algorithm contains the following line:

     If not($x > 5$ and $x \leq 10$) **then** …

How could this be written more simply?

SOLUTION

Apply the first de Morgan's law: $\neg[(x > 5) \wedge (x \leq 10)]$ is equivalent to $\neg(x > 5) \vee \neg(x \leq 10)$, which in turn is equivalent to $(x \leq 5) \vee (x > 10)$. The line of the algorithm can therefore be written:

     If $x \leq 5$ or $x > 10$ **then** …

The next example shows how a type of problem that we previously dealt with using truth tables can also be solved using the laws of logic.

EXAMPLE 4.6.4

Use the laws of logic to show that $[(p \rightarrow q) \wedge \neg q] \rightarrow \neg p$ is a tautology.

SOLUTION

$[(p \rightarrow q) \wedge \neg q] \rightarrow \neg p \equiv \neg[(\neg p \vee q)] \wedge \neg q] \vee \neg p$	implication law (twice)
$\equiv \neg[\neg q \wedge (\neg p \vee q)] \vee \neg p$	first commutative law
$\equiv \neg[(\neg q \wedge \neg p) \vee (\neg q \wedge q)] \vee \neg p$	first distributive law
$\equiv \neg[(\neg q \wedge \neg p) \vee (q \wedge \neg q)] \vee \neg p$	first commutative law
$\equiv \neg[(\neg q \wedge \neg p) \vee F] \vee \neg p$	first inverse law
$\equiv \neg(\neg q \wedge \neg p) \vee \neg p$	second identity law
$\equiv (\neg\neg q \vee \neg\neg p) \vee \neg p$	first de Morgan's law
$\equiv (q \vee p) \vee \neg p$	double negation law (twice)
$\equiv q \vee (p \vee \neg p)$	second associative law
$\equiv q \vee T$	second inverse law
$\equiv T$	second annihilation law

Therefore $[(p \rightarrow q) \wedge \neg q] \rightarrow \neg p$ is a tautology.

Both methods have advantages and disadvantages. The truth table method can be lengthy, but it is a mechanical procedure guaranteed to lead to the answer eventually. Applying the laws of logic can be more difficult, because it is not always easy to decide which law should be applied in a given situation. On the other hand, the laws of logic will often lead to a solution more quickly.

The next example shows how the laws of logic can be used to determine the validity of an argument.

EXAMPLE 4.6.5

Determine whether the following argument is valid:

'The file is either a binary file or a text file. If it is a binary file then my program won't accept it. My program will accept the file. Therefore the file is a text file.'

SOLUTION

An argument of this type consists of some *premises* (in this example, the first three sentences), which together are supposed to imply the *conclusion* (the last sentence). The argument takes the form of the logical expression:

$$(P_1 \wedge P_2 \wedge P_3) \rightarrow Q$$

where P_1, P_2 and P_3 are the premises, and Q is the conclusion. (There is no ambiguity in writing $P_1 \wedge P_2 \wedge P_3$ without brackets, because the connective $\wedge$ obeys the associative law.) If the argument is valid, the expression should be a tautology.

Let p denote the proposition 'The file is a binary file', let q denote 'The file is a text file', and let r denote 'My program will accept the file'. Then:

$$P_1 \equiv p \vee q$$
$$P_2 \equiv p \rightarrow \neg r$$
$$P_3 \equiv r$$
$$Q \equiv q$$

The argument now takes the form:

$$[(p \vee q) \wedge (p \rightarrow \neg r) \wedge r] \rightarrow q$$

We can find out whether this expression is a tautology either by constructing a truth table or by trying to simplify it using the laws of logic. Since a truth table would require eight rows and a large number of columns, and would be fairly tedious to construct, we try the latter approach.

$$
\begin{aligned}
[(p \vee q) \wedge (p \rightarrow \neg r) \wedge r] \rightarrow q &\equiv \neg[(p \vee q) \wedge (\neg p \vee \neg r) \wedge r] \vee q \\
&\equiv \neg[(p \vee q) \wedge r \wedge (\neg p \vee \neg r)] \vee q \\
&\equiv \neg\{(p \vee q) \wedge [(r \wedge \neg p) \vee (r \wedge \neg r)]\} \vee q \\
&\equiv \neg\{(p \vee q) \wedge [(r \wedge \neg p) \vee F]\} \vee q \\
&\equiv \neg[(p \vee q) \wedge r \wedge \neg p] \vee q \\
&\equiv \neg[\neg p \wedge (p \vee q) \wedge r] \vee q \\
&\equiv \neg\{[(\neg p \wedge p) \vee (\neg p \wedge q)] \wedge r\} \vee q \\
&\equiv \neg\{[F \vee (\neg p \wedge q)] \wedge r\} \vee q \\
&\equiv \neg(\neg p \wedge q \wedge r) \vee q \\
&\equiv \neg(q \wedge \neg p \wedge r) \vee q \\
&\equiv \neg q \vee \neg(\neg p \wedge r) \vee q \\
&\equiv \neg q \vee q \vee \neg(\neg p \wedge r) \\
&\equiv T \vee \neg(\neg p \wedge r) \\
&\equiv T
\end{aligned}
$$

The proposition is a tautology, so the argument is valid.

4.7 *Predicate logic*

In the final section of this chapter, we take a brief look at predicate logic.

Propositional logic provides a useful setting in which we can analyse many types of logical argument. There are situations, however, where propositional logic is inadequate, because it cannot deal with the logical structure that is sometimes present *within* atomic propositions.

Consider the following arguments:

'All even numbers are integers. 8 is an even number. Therefore 8 is an integer.'

'It is not true that all prime numbers are odd. Therefore there must be at least one prime number that is not odd.'

Both of these arguments appear to be perfectly valid on the basis of everyday reasoning, yet if we try to show their validity using propositional logic we run into difficulties.

In the first argument, the atomic propositions are 'All even numbers are integers', '8 is an even number', and '8 is an integer'. The argument takes the following form:

$$(p \wedge q) \to r$$

This expression is false if p and q are true and r is false, so it is not a tautology.

Similarly, if we take the atomic propositions in the second argument to be 'All prime numbers are odd' and 'There must be at least one prime number that is not odd', the argument takes the form:

$$\neg p \to q$$

This expression is also not a tautology, because it is false if p and q are both false.

In order to be able to analyse arguments such as these, we need to look at the logical structure within atomic propositions. Predicate logic allows us to do this.

Definition

A *predicate* is a statement containing one or more variables. If values are assigned to all the variables in a predicate, the resulting statement is a proposition.

For example, '$x < 5$' is a predicate, where x is a variable denoting any real number. If we substitute a real number for x, we obtain a proposition; for example, '$3 < 5$' and '$6 < 5$' are propositions with truth values T and F respectively.

A variable need not be a number. For example, 'x is an employee of the Ezisoft Software Company' becomes a proposition with a well defined truth value when x is replaced by a person's name: 'Frederick Firestone[2] is an employee of the Ezisoft Software Company'.

There are other ways of obtaining a proposition from a predicate apart from assigning values to the variables. For example, consider the predicate '$x < 5$ or $x \geq 5$'. This predicate is true no matter what value we substitute for x, so we can form a true proposition by writing:

'For all x, $x < 5$ or $x \geq 5$.'

---

[2]The name 'Frederick Firestone' appealed to the author when he saw it on a freeway exit sign in Colorado. (The sign actually marks the exit to two towns called Frederick and Firestone.)

If we had used '$x < 5$' as the predicate instead of '$x < 5$ or $x \geq 5$', we would have obtained:

'For all x, $x < 5$',

which is also a proposition (albeit a false one).

While the predicate '$x < 5$' is not always true, it *is* true for some values of x, so we can form a true proposition by writing:

'There exists an x such that $x < 5$.' (Here, 'an' means 'at least one'.)

The expressions 'for all' and 'there exists' are called *quantifiers*. The process of applying a quantifier to a variable is called *quantifying* the variable. A variable that has been quantified is said to be *bound*. For example, the variable x in 'There exists an x such that $x < 5$' is bound by the quantifier 'there exists'. A variable that appears in a predicate but is not bound is said to be *free*.

We now want to introduce a notation that will allow us to write predicates and quantifiers symbolically. We will use capital letters to denote predicates. A predicate P that contains a variable x can be written symbolically as $P(x)$. A predicate can contain more than one variable; a predicate P with two variables, x and y for example, can be written $P(x, y)$. In general, a predicate with n variables, $x_1, x_2, \ldots, x_n$, can be written $P(x_1, x_2, \ldots, x_n)$.

The quantifiers 'for all' and 'there exists' are denoted by the symbols $\forall$ and $\exists$ respectively. With this notation, expressions containing predicates and quantifiers can be written symbolically.

EXAMPLE **4.7.1**

Write in symbols: 'There exists an x such that $x < 4$.'

SOLUTION

Let $P(x)$ be '$x < 4$'. Then the proposition can be written:

$$\exists x \, P(x)$$

EXAMPLE **4.7.2**

Write in symbols: 'For all x, $x < 5$ or $x \geq 5$'.

SOLUTION

Let $P(x)$ be '$x < 5$', and let $Q(x)$ be '$x \geq 5$'. Then the proposition can be written:

$$\forall x \, [P(x) \lor Q(x)]$$

If we use the fact that $Q(x)$ is equivalent to $\neg P(x)$, we can also write:

$$\forall x \, [P(x) \vee \neg P(x)]$$

Here is a more complicated example, using a predicate with two variables.

EXAMPLE 4.7.3

Write the following two propositions in symbols:

'For every number x there is a number y such that $y = x + 1$'.

'There is a number y such that, for every number x, $y = x + 1$'.

SOLUTION

Let $P(x, y)$ denote the predicate '$y = x + 1$'. The first proposition is:

$$\forall x \, \exists y \, P(x, y)$$

The second proposition is:

$$\exists y \, \forall x \, P(x, y)$$

Note carefully the difference in meaning between the two propositions in Example 4.7.3. In the first proposition the value of y can depend on x (i.e. different values of x can give different values of y), whereas in the second proposition it cannot. In fact, the first proposition is a true statement about numbers, while the second is a false statement. This example shows that the order in which the quantifiers appear can affect the meaning.

The following example suggests a practical problem in which the notation of predicate logic is useful.

EXAMPLE 4.7.4

In the specification of a system for booking theatre seats, $B(p, s)$ denotes the predicate 'person p has booked seat s'. Write the following sentences in symbolic form:

(a) Seat s has been booked.
(b) Person p has booked a (i.e. at least one) seat.
(c) All the seats are booked.
(d) No seat is booked by more than one person.

SOLUTION

(a) $\exists p \; B(p, s)$

(b) $\exists s \; B(p, s)$

(c) $\forall s \; \exists p \; B(p, s)$

(d) If no seat is booked by more than one person, then $B(p, s)$ and $B(q, s)$ cannot both be true unless p and q denote the same person. In symbols:

$$\forall s \; \forall p \; \forall q \; \{[B(p, s) \wedge B(q, s)] \to (p = q)\}$$

It would be possible to formulate various laws of logic involving the two quantifiers $\forall$ and $\exists$ and the connectives we introduced earlier. Here we will look just at the relationship between the two quantifiers and the connective **not**.

Suppose we want to apply the connective **not** to the following proposition:

'All swans are black.'

Applying **not** to a proposition is called *negating* the proposition.

The original proposition can be written in symbols:

$$\forall x \; P(x)$$

where $P(x)$ is the predicate 'Swan x is black'.

Here is one way of forming the negation:

'It is not true that all swans are black.'

Or, more simply:

'Not all swans are black.'

We can write this proposition in symbols as follows:

$$\neg[\forall x \; P(x)]$$

Note that it would be incorrect to give the negation as 'All swans are not black'. This would be saying something different — that there are *no* black swans.

There is another way of saying that not all swans are black; we can say that *there must be at least one swan that is not black*. This gives us an alternative way of expressing the negation of the original proposition:

'There is a swan that is not black.'

In symbols:

$$\exists x \; [\neg P(x)]$$

By comparing this form of the negation with the original proposition 'All swans are black', we can see that forming the negation corresponds to *negating the predicate and changing the quantifier*. We can express this observation as a law of predicate logic:

$$\neg[\forall x \; P(x)] \equiv \exists x \; [\neg P(x)]$$

There is a second law, which can be thought of as the dual of the first, for negating a proposition containing 'there exists'. It also corresponds to negating the predicate and changing the quantifier:

$$\neg[\exists x\ P(x)] \equiv \forall x\ [\neg P(x)]$$

For example, the negation of 'There is a number x such that $x^2 = 2$' is 'For every number x, $x^2 \neq 2$'.

EXAMPLE 4.7.5

Write down the negation of the following proposition:

'For every number x there is a number y such that $y < x$.'

SOLUTION

Write the negation in symbols and simplify it using the laws of logic:

$$\begin{aligned}
\neg[\forall x\ \exists y\ (y < x)] &\equiv \exists x\ \{\neg[\exists y\ (y < x)]\} \\
&\equiv \exists x\ \forall y\ [\neg(y < x)] \\
&\equiv \exists x\ \forall y\ (y \geq x)
\end{aligned}$$

Write the answer as an English sentence:

'There is a number x such that, for every number y, $y \geq x$.'

It is important in an example like this to check that the answer makes sense in terms of what the proposition and its negation mean, rather than just mechanically applying the laws of logic. In Example 4.7.5, the original proposition is a true mathematical statement about real numbers, while its negation is a false statement.

The rule for negating the quantifier 'for all' can be used to verify one of the arguments we quoted as an example at the beginning of this section:

'It is not true that all prime numbers are odd. Therefore there must be at least one prime number that is not odd.'

Let $P(x)$ denote the predicate 'x is a prime number', and let $Q(x)$ denote the predicate 'x is odd'. The proposition 'all prime numbers are odd' can be rephrased as 'for all x, if x is a prime number then x is odd', and written in symbolic form in the following way:

$$\forall x\ [P(x) \to Q(x)]$$

Therefore the first sentence of the argument can be written in symbols as follows:

$$\neg\{\forall x\ [P(x) \to Q(x)]\}$$

If we apply the negation rule and the laws of propositional logic we met in Section 4.6, we obtain the following equivalent expressions:

$$\neg\{\forall x\,[P(x) \to Q(x)]\} \equiv \exists x\,\{\neg[P(x) \to Q(x)]\}$$
$$\equiv \exists x\,\{\neg[\neg P(x) \vee Q(x)]\}$$
$$\equiv \exists x\,[\neg\neg P(x) \wedge \neg Q(x)]$$
$$\equiv \exists x\,[P(x) \wedge \neg Q(x)]$$

The last line reads: 'There is a number x such that x is a prime number and x is not odd', which we can rephrase as: 'There must be at least one prime number that is not odd'.

Many of the ideas we have met in this chapter play a fundamental role in computing and mathematics, and they will reappear frequently in one form or another in the following chapters. In particular, we will see in Chapter 8 how the laws of logic (in the guise of Boolean algebra) can be used to study the design of the circuitry on which modern digital computers are based.

▶ *Exercises* ▶

1. Express the following propositions in symbolic form, and identify the principal connective:

   (a) Either Karen is studying computing and Minh is not studying mathematics, or Minh is studying mathematics.

   (b) It is not the case that if it is sunny then I will carry an umbrella.

   (c) The program will terminate if and only if the input is not numeric or the escape key is pressed.

   (d) If $x = 7$ and $y \neq 4$ and $z = 2$, then if it is not true that either $y = 4$ or $z \neq 2$ then $x = 7$ or $z = 2$.

   (Assume that this sentence arises in a context in which x, y and z have been assigned values, so that it is a genuine proposition.)

2. Let p and q denote respectively the propositions 'It is snowing' and 'I will go skiing'. Write down English sentences corresponding to the following propositions:

   (a) $\neg p \wedge q$

   (b) $p \to q$

   (c) $\neg q \to p$

   (d) $(p \vee \neg q) \wedge p$

3. (a) Construct the truth table for the connective **xor** with symbol $\oplus$, where $p \oplus q$ means 'either p or q but not both'.

   (b) Construct a truth table to show that $p \oplus q$ is logically equivalent to $(p \vee q) \wedge \neg(p \wedge q)$.

4. Write down English sentences for the converse and contrapositive of the following propositions:

   (a) If the input file exists, then an error message is not generated.

(b) If a graphics driver is not available, then my program cannot run.

(c) If my program contains no bugs, then it produces correct output.

5. Write down English sentences corresponding to the converse and the contrapositive of $p \rightarrow q$, where p and q are defined in Exercise 2.

6. Construct truth tables for the following expressions. In each case, state whether the expression is a tautology, a contradiction, or neither.

(a) $\neg(p \vee \neg q) \vee p$

(b) $[p \rightarrow (p \wedge q)] \rightarrow \neg q$

(c) $(p \wedge q) \leftrightarrow (\neg p \vee \neg q)$

(d) $[(p \wedge r) \vee (q \wedge r)] \rightarrow (p \rightarrow \neg q)$

7. Let P and Q denote two logical expressions. If P is false for a particular set of truth values of the variables, then $P \wedge Q$ must be false for that set of values, so there is no need to find the truth value of Q.

(a) State a similar rule involving $P \vee Q$.

(b) Using these two rules as short-cuts, construct the truth tables for the following expressions. (The rules mean that some of the entries in the table may be left blank, but the last column must still be complete.)

(i) $[\neg(p \wedge q) \wedge (p \vee \neg r)] \wedge [(p \wedge r) \vee \neg q]$

(ii) $\neg[\neg p \wedge (q \vee r)] \vee (\neg p \wedge \neg r)$

8. Use truth tables to show that $\neg(p \vee \neg q)$ and $\neg p \wedge q$ are logically equivalent.

9. Using truth tables, prove the following laws of logic:

(a) $p \wedge (q \vee r) \equiv (p \wedge q) \vee (p \wedge r)$

(b) $p \wedge (p \vee q) \equiv p$

10. Use the laws of logic to simplify the following expressions as far as possible:

(a) $(p \vee \neg q) \wedge (p \vee q)$

(b) $\neg[p \rightarrow \neg(p \wedge q)]$

(c) $\neg[p \vee (q \wedge \neg p)]$

(d) $[(p \leftrightarrow q) \rightarrow \neg(r \rightarrow p)] \vee (r \rightarrow \neg q)$

11. An algorithm contains the following line:

If not($x \geq 3$ and $x < 6$) **then** …

How could this be written more simply?

12. Rewrite the following pseudocode using a **While-do** in place of the **Repeat-until**:

1. $n \leftarrow 0$
2. *term* $\leftarrow 1$
3. *sum* $\leftarrow 0$
4. **Repeat**
    4.1. $n \leftarrow n + 1$
    4.2. *term* $\leftarrow$ *term* / 2
    4.3. *sum* $\leftarrow$ *sum* + *term*
    until *term* < 0.001 or $n = 100$

13. Use the laws of logic to classify the following expressions as tautologies or contradictions:

    (a) $(p \wedge \neg q) \vee (\neg p \vee q)$
    (b) $[p \rightarrow (q \rightarrow p)] \leftrightarrow (p \wedge \neg p)$
    (c) $[p \wedge (p \rightarrow q)] \rightarrow q$

14. Express the following argument in symbolic form and test its validity using the laws of logic:

    'If $n > 10$ when the subroutine call statement is reached, then the subroutine is called. The subroutine is called. Therefore $n > 10$ when the subroutine call statement is reached.'

15. Express the following argument in symbolic form and test its validity using the laws of logic:

    'Sandra is studying computing or she is not studying accounting. If Sandra is studying accounting then she is not studying computing. Therefore Sandra is studying computing.'

16. Find an expression that is logically equivalent to $p \vee q$ but uses only the connectives **and** and **not**.

17. The connective **nand**, with symbol $|$ (sometimes called the *Sheffer stroke*), is defined by the truth table shown in Table 4.14.

p	q	$p \mid q$
T	T	F
T	F	T
F	T	T
F	F	T

Table 4.14

    (a) Find an expression that is logically equivalent to $\neg p$ using only the connective **nand**.
    (b) Find an expression that is logically equivalent to $p \wedge q$ using only the connective **nand**.
    (c) Find an expression that is logically equivalent to $p \vee q$ using only the connective **nand**.

    (This exercise shows that any expression built up using the connectives **and, or** and **not** can be converted to a logically equivalent expression using just the connective **nand**.)

18. Write the following propositions symbolically in the notation of predicate logic, and state their truth values:

    (a) 'There is a real number x such that $x^2 - 3x + 2 = 0$.'
    (b) 'For every real number x there is a real number y such that $x = y^2$.'

19. Write the negations of the propositions in Exercise 18 in symbolic form and in English.

20. In the design specification of a library borrowing system, $B(p, b)$ denotes the predicate 'person p has borrowed book b', and $O(b)$ denotes the predicate 'book b is overdue'.

Write the following sentences in symbolic form:

(a) Person p has borrowed a book. (Assume that 'a' means 'at least one'.)
(b) Book b has been borrowed.
(c) Book b is on the shelf.
(d) Person p has borrowed at least two books.
(e) No book has been borrowed by more than one person.
(f) There are no overdue books.
(g) If a book is overdue, then it must have been borrowed.
(h) Person p has an overdue book.

21. Consider the following self-referential statement:

'This statement has five words.'

(a) What is the truth value of the statement?
(b) Write down the negation of the statement. What is its truth value?

(This exercise shows the kind of difficulty that can arise with a self-referential statement, even if it appears to have a well defined truth value.)

22. On one side of a card is written:

'The statement on the other side of this card is true.'

On the other side of the card is written:

'The statement on the other side of this card is false.'

Explain how a paradox arises in this situation. (The problem is known as the Jourdain card paradox.)

23. Four people are using computers in a computing laboratory. You know that the first person is a student and the second is not, but you do not know whether they are using the software on the network. You know that the third person is using the software on the network and the fourth is not, but you do not know whether they are students.

As the laboratory supervisor, you are required to enforce the rule that only students are allowed to use the software on the network. Which two people should you question, and what should you ask them?

5. Sets and Relations

5.1 Sets

In this chapter, we will present some basic mathematical ideas about sets and relations. Some of the material on sets may be familiar to you already, in which case you may wish to scan over those sections fairly briefly. The main reason for introducing sets is to provide some useful terminology and notation for the work that follows; we will not be studying the mathematical theory of sets as such. Relations arise in computing in the area of relational databases, and we will also need them in Chapter 12 when we study congruences.

The word *set* is used in mathematics to mean any well defined collection of items. The items in a set are called the *elements* of the set. For example, we can refer to the set of all the employees of a particular company, the set of all ASCII characters[1], or the set of all the integers that are divisible by 5.

A specific set can be defined in two ways. If there are only a few elements, they can be listed individually, by writing them between braces ('curly' brackets) and placing commas in between. For example, the set of positive odd numbers less than 10 can be written in the following way:

$$\{1, 3, 5, 7, 9\}$$

If there is a clear pattern to the elements, an ellipsis (three dots) can be used. For example, the set of odd numbers between 0 and 50 can be written:

$$\{1, 3, 5, 7, ..., 49\}$$

Some infinite sets can also be written in this way; for example, the set of all positive odd numbers can be written:

$$\{1, 3, 5, 7, ...\}$$

A set written in any of these ways is said to be written in *enumerated form*.

The second way of writing down a set is to use a property that defines the elements of the set. Braces are used with this notation also. For example, the set of odd numbers between 0 and 50 can be written:

$$\{x: x \text{ is odd and } 0 < x < 50\}$$

---

[1]The ASCII character set is a standard set of 128 characters, including letters, digits, punctuation marks, mathematical symbols and non-printing (control) characters. Each character has a unique numeric code from 0 to 127. The 8-bit binary equivalent of the numeric code is used in many computers for the internal representation of character data. (ASCII stands for American Standard Code for Information Interchange.)

The colon is read 'such that', so the definition reads 'the set of all x such that x is odd and $0 < x < 50$'. Recalling our work from the previous chapter, notice that the expression following the colon is a predicate containing the variable x. A set written in the form $\{x: P(x)\}$, where $P(x)$ is a predicate, is said to be written in *predicate form*.

Capital letters are commonly used to denote sets. For example, we can write:

$$A = \{1, 2, 3, 4, 5\}$$

$$B = \{x: x \text{ is a multiple of } 3\}$$

The symbol $\in$ means 'is an element of'. For example, if A and B are the two sets defined above, we can write $2 \in A$ and $15 \in B$. The symbol $\notin$ means 'is not an element of'; for example, $6 \notin A$ and $11 \notin B$.

Many (but by no means all) of the sets we will be dealing with are sets of numbers. Some sets of numbers arise sufficiently often that special symbols are reserved for them. The most important of these for our subsequent work are listed below:

- **N** is the set of natural numbers (or positive integers): $\{1, 2, 3, 4, \ldots\}$.
- **J** is the set of integers: $\{\ldots, -3, -2, -1, 0, 1, 2, 3, \ldots\}$.
- **Q** is the set of rational numbers: $\{x: x = m/n \text{ for some integers } m \text{ and } n\}$.
- **R** is the set of real numbers.

The sets **J** and **R** may remind you of the data types 'integer' and 'real', which arise in programming. We will not go into the details of the concept of typing here. However, it is important to point out that a careful distinction needs to be made between the elements of **J** and **R** and the representations of integers and real numbers in a computer. Data types in a computer are constrained by the technical limitations of computing hardware; we saw in Chapter 3 that there is a limit to the size (and, in the case of real numbers, the accuracy) of the numbers that can be represented. There is no such restriction when we are doing mathematics; we are quite free to imagine infinite sets of numbers, and it is often convenient to do so.

There are two more sets for which we will introduce special symbols. The first of these is the *null* set (or empty set), which is denoted by the symbol $\emptyset$, and which has no elements. (Note that the symbol $\emptyset$ is *not* the same as the Greek letter ϕ (phi).) The null set can be written in enumerated form, like this:

$$\{\}$$

Don't make the mistake of writing $\{\emptyset\}$ for the null set. The set $\{\emptyset\}$ is *not* the null set; it is a set with one element, and that element is the null set.

The null set can be written in predicate form using any predicate that is always false, for example $\{x: x \neq x\}$.

It may seem strange to define the null set, and you might wonder why it is necessary to do so. Perhaps you had similar misgivings when you first learnt about the number 0 ('If there's nothing there, how can you count it?'). It will become clear as we proceed that we need the null set in our work.

The other set for which we will introduce a symbol is the *universal* set, denoted by $\mathscr{E}$. The term 'universal set' does not refer to a specific set, but rather to a set that contains all the elements arising in the problem at hand. The universal set can therefore change from one problem to another. For example, in a problem dealing with various sets of numbers, we might choose $\mathscr{E}$ to be **R**, the set of all real numbers.

5.2 Subsets, set operations and Venn diagrams

Definition

Let *A* and *B* be sets. We say that *B* is a *subset* of *A*, and write $B \subseteq A$, if every element of *B* is an element of *A*.

For example, let $A = \{1, 2, 3, 4, 5\}$, $B = \{1, 3, 4\}$, and $C = \{2, 4, 6\}$. Then $B \subseteq A$, but *C* is not a subset of *A*, because $6 \in C$ but $6 \notin A$.

A useful way of depicting the relationship between several sets is to represent each set by an oval region on a type of diagram known as a *Venn*[2] *diagram*. The Venn diagram for the present example is shown in Figure 5.1.

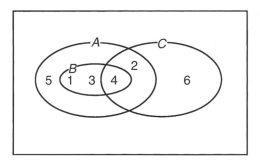

Figure 5.1

The rectangle forming the boundary of the Venn diagram represents the universal set $\mathscr{E}$, which we may take to be $\{1, 2, 3, 4, 5, 6\}$ in this example.

---

[2]Venn diagrams are named after the English mathematician, John Venn (1834–1923).

A second example of subsets is provided by the following 'chain' of subsets of **R**, the set of real numbers:

$$\mathbf{N} \subseteq \mathbf{J} \subseteq \mathbf{Q} \subseteq \mathbf{R}$$

The definition of 'subset' should be read carefully (as should all definitions in mathematics); it really does mean *exactly* what it says. In particular, if A is any set, then $A \subseteq A$, because it is certainly true that every element of A is an element of A. (The resemblance of the symbol $\subseteq$ to $\leq$ is not accidental; the relations 'subset' and 'less than or equal to' have some similar properties.) If $B \subseteq A$ and $B \neq A$, then B is called a *proper* subset of A.

The reasoning that underlies the following fact is a little more subtle:

$$\varnothing \subseteq A \text{ for any set } A$$

In words: *every element of $\varnothing$ is an element of A.* If you are puzzled by this statement, ask yourself how one set can *fail* to be a subset of another. This can happen only if there is an element of the first set that is not an element of the second; for example, $\{2, 4, 6\}$ is not a subset of $\{1, 2, 3, 4, 5\}$, because 6 is an element of the first set but not the second. Therefore the statement $\varnothing \subseteq A$ would be false *only* if we could find an element of $\varnothing$ that is not an element of A. But we can't find such an element, because $\varnothing$ has no elements! We say that the statement 'every element of $\varnothing$ is an element of A' is *vacuously* true, and conclude that $\varnothing \subseteq A$.

Now that we have established the concept of a subset of a set, we can define what it means for two sets to be equal.

Definition

Two sets A and B are *equal* if $A \subseteq B$ and $B \subseteq A$.

In other words, $A = B$ if every element of A is an element of B, and every element of B is an element of A. A less formal way of expressing this is: 'Two sets are equal if they have the same elements'. In particular, this means that the order in which the elements of a set are listed in enumerated form is unimportant: $\{a, b, c\}$ is the same set as $\{b, c, a\}$. It also means that a set does not have 'repeated' elements — we would never write a set as $\{a, a, b\}$, because it would be the same set as $\{a, b\}$.

A number of operations are defined on sets. We list them now, together with their corresponding Venn diagrams.

- The *intersection* of two sets A and B is:

$$A \cap B = \{x: x \in A \text{ and } x \in B\}$$

The intersection of A and B is depicted by the shaded region of the Venn diagram in Figure 5.2.

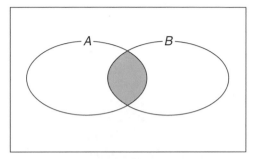

Figure 5.2

Two sets with no elements in common are said to be *disjoint*. In symbols, A and B are disjoint if $A \cap B = \varnothing$.

- The *union* of A and B is:

$$A \cup B = \{x: x \in A \text{ or } x \in B\}$$

(Recall that 'or' always means inclusive-or.)

The union of A and B is depicted by the shaded region of the Venn diagram in Figure 5.3.

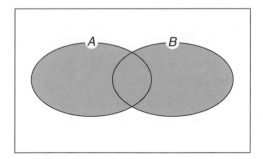

Figure 5.3

- The *complement* of A is:

$$\overline{A} = \{x: x \in \mathscr{E} \text{ and } x \notin A\}$$

(Recall that $\mathscr{E}$ is the universal set.)

The complement of *A* is depicted by the shaded region of the Venn diagram in Figure 5.4.

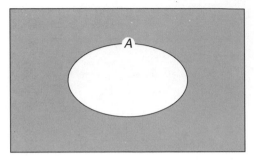

Figure 5.4

- The *difference* of *A* and *B* is:

$$A - B = \{x: x \in A \text{ and } x \notin B\}$$

The difference of *A* and *B* is depicted by the shaded region of the Venn diagram in Figure 5.5.

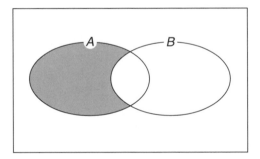

Figure 5.5

The last of the above operations is less widely used than the others, because it can always be rewritten in terms of intersection and complement, using the law:

$$A - B = A \cap \overline{B}$$

We can illustrate a law such as this one by constructing the Venn diagram for each side of the equation, and seeing that the two diagrams are the same. In this case, the Venn diagram for the right-hand side is produced by shading the regions corresponding to *A* and $\overline{B}$, as shown in Figure 5.6. Then $A \cap \overline{B}$ corresponds to the doubly shaded region (indicated by darker shading in the figure). This region is the same as the region shaded in Figure 5.5.

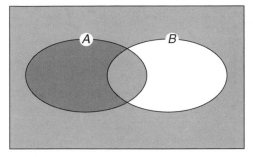

Figure 5.6

Alternatively, we can prove that $A - B = A \cap \overline{B}$ using the definitions of the set operations:

$$A - B = \{x: x \in A \text{ and } x \notin B\} = \{x: x \in A\} \cap \{x: x \notin B\} = A \cap \overline{B}$$

EXAMPLE 5.2.1

Let the universal set be $\mathcal{E} = \{1, 2, 3, \dots, 10\}$. Let $A = \{2, 4, 7, 9\}$, $B = \{1, 4, 6, 7, 10\}$, and $C = \{3, 5, 7, 9\}$. Find:

(a) $A \cup B$
(b) $A \cap C$
(c) $B \cap \overline{C}$
(d) $(A \cap \overline{B}) \cup C$
(e) $\overline{B \cup C} \cap C$

SOLUTION

It is helpful to draw a Venn diagram in a problem like this (Figure 5.7).

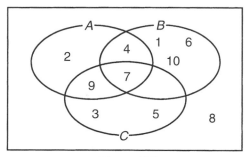

Figure 5.7

(a) $A \cup B = \{1, 2, 4, 6, 7, 9, 10\}$
(b) $A \cap C = \{7, 9\}$
(c) $\overline{C} = \{1, 2, 4, 6, 8, 10\}$, so $B \cap \overline{C} = \{1, 4, 6, 10\}$

(d) $\overline{B} = \{2, 3, 5, 8, 9\}$, therefore $A \cap \overline{B} = \{2, 9\}$, so $(A \cap \overline{B}) \cup C = \{2, 3, 5, 7, 9\}$

(e) $B \cup C = \{1, 3, 4, 5, 6, 7, 9, 10\}$, therefore $\overline{B \cup C} = \{2, 8\}$, so $\overline{B \cup C} \cap C = \varnothing$

The operations of intersection, union and complement correspond in a natural way to the logical connectives **and**, **or** and **not** respectively. Because of this, each of the laws of logic gives rise to a corresponding law of sets.

For example, the commutative law $p \wedge q \equiv q \wedge p$ in logic yields the law $A \cap B = B \cap A$ for sets. We can show this in the following way:

$$A \cap B = \{x: x \in A \text{ and } x \in B\} = \{x: x \in B \text{ and } x \in A\} = B \cap A$$

The first and third equalities follow from the definition of the intersection of two sets, while the second equality follows from the commutative law of logic.

A list of some of the laws of sets is given in Table 5.1. Most of these laws have the same name as the corresponding law of logic, and all can be derived from one of the laws of logic in the way we have just seen.

Law(s)		Name
	$\overline{\overline{A}} = A$	double complement law
$A \cap A = A$	$A \cup A = A$	idempotent laws
$A \cap B = B \cap A$	$A \cup B = B \cup A$	commutative laws
$(A \cap B) \cap C = A \cap (B \cap C)$	$(A \cup B) \cup C = A \cup (B \cup C)$	associative laws
$A \cap (B \cup C) =$ $(A \cap B) \cup (A \cap C)$	$A \cup (B \cap C) =$ $(A \cup B) \cap (A \cup C)$	distributive laws
$\overline{A \cap B} = \overline{A} \cup \overline{B}$	$\overline{A \cup B} = \overline{A} \cap \overline{B}$	de Morgan's laws
$A \cap \mathscr{E} = A$	$A \cup \varnothing = A$	identity laws
$A \cap \varnothing = \varnothing$	$A \cup \mathscr{E} = \mathscr{E}$	annihilation laws
$A \cap \overline{A} = \varnothing$	$A \cup \overline{A} = \mathscr{E}$	inverse laws
$A \cap (A \cup B) = A$	$A \cup (A \cap B) = A$	absorption laws

Table 5.1

EXAMPLE 5.2.2

Use Venn diagrams to illustrate the first de Morgan's law for sets.

SOLUTION

We draw a Venn diagram for each side of the equation. For the left-hand side, we draw the Venn diagram for $A \cap B$ first, and then draw the Venn diagram for $\overline{A \cap B}$ (Figure 5.8).

Figure 5.8

For the right-hand side, we draw the Venn diagrams shown in Figure 5.9.

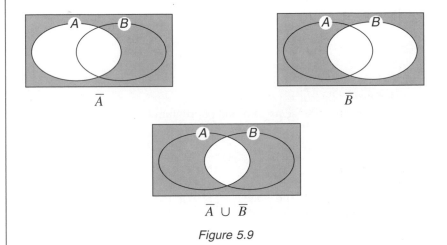

Figure 5.9

The shaded region is the same for both sides of the equation.

5.3 *Cardinality and Cartesian products*

The *cardinality* of a finite[3] set is the number of elements in the set. The cardinality of a set A is denoted by $|A|$.

For example, if $A = \{a, b, c, d, e\}$ then $|A| = 5$.

The following formula for the cardinality of the union of two sets is known as the Principle of inclusion and exclusion:

$$|A \cup B| = |A| + |B| - |A \cap B|$$

The formula is derived from the fact that if the cardinalities of A and B are added, then the elements in $A \cap B$ will be counted twice, so this is corrected for by subtracting the cardinality of $A \cap B$.

---

[3]It is possible to define cardinality for infinite sets also, but we will not do this here.

EXAMPLE 5.3.1

In a group of 120 students studying computing, 84 can program in Pascal and 66 can program in Lisp. If 45 can program in both Pascal and Lisp, how many of the students cannot program in either of these languages?

SOLUTION

Let $\mathcal{E}$ = {computing students}, P = {Pascal programming students}, and L = {Lisp programming students}. The problem is to find $|\overline{P \cup L}|$. By the Principle of inclusion and exclusion:

$$|P \cup L| = |P| + |L| - |P \cap L|$$
$$= 84 + 66 - 45$$
$$= 105$$

Therefore $|\overline{P \cup L}| = 120 - 105 = 15$.

There are 15 students who cannot program in either of these languages.

Definition

Let A be a set. The *power set* of A is the set of all subsets of A, and is denoted by $\mathcal{P}(A)$.

Note that a power set is an example of a *set of sets*, that is, a set whose elements are themselves sets.

For example, consider a set with 3 elements: {a, b, c}. Listing all the subsets of {a, b, c}, and remembering to include the set itself and the null set, we obtain the following power set:

$\mathcal{P}(\{a, b, c\}) = \{\varnothing, \{a\}, \{b\}, \{c\}, \{a, b\}, \{a, c\}, \{b, c\}, \{a, b, c\}\}$

The power set of {a, b, c} has 8 elements. Since {a, b, c} can represent any 3-element set, we conclude that any set with 3 elements has 8 subsets.

A general formula for the number of subsets of any finite set is given by the following result.

Theorem

Let A be a set with n elements. Then A has 2^n subsets.

Proof
Let $A = \{x_1, x_2, ..., x_n\}$, and suppose we want to select a subset B of A. We can do this by looking at each element of A in turn, and deciding whether or not to include it in B. There are two possibilities for x_1: either $x_1 \in B$ or $x_1 \notin B$. Similarly, either $x_2 \in B$ or $x_2 \notin B$ (two possibilities), so the total number of possibilities for x_1 and x_2 is 2×2. Continuing in this way with $x_3, x_4, ..., x_n$, we conclude that the total number of possible subsets is $2 \times 2 \times ... \times 2$ (n times), which equals 2^n.

The technique used in the above proof is an application of a result known as the Multiplication principle, which we will meet again in Chapter 9.

We noted earlier that the elements of a set are not listed in any particular order — {a, b} is the same set as {b, a}, for example. By contrast, an *ordered n-tuple* is a list of n elements *in a particular order*. To distinguish them from sets, ordered *n*-tuples are written using parentheses instead of braces:

$$(x_1, x_2, ..., x_n)$$

In particular, an ordered 2-tuple (usually called an *ordered pair*) is a pair of elements in a particular order, and is written (a, b). Thus the ordered pair $(1, 2)$ is different from $(2, 1)$, for example.

If two ordered *n*-tuples $(x_1, x_2, ..., x_n)$ and $(y_1, y_2, ..., y_n)$ are equal, then the elements in the corresponding positions must be equal: $x_1 = y_1$, $x_2 = y_2$, and so on.

Definition

The *Cartesian*[4] *product* of two sets A and B is defined by:

$$A \times B = \{(x, y): x \in A \text{ and } y \in B\}$$

In words: $A \times B$ is the set of all ordered pairs in which the first element comes from A and the second element comes from B.

More generally, the Cartesian product of the n sets $A_1, A_2, ..., A_n$ is defined by:

$$A_1 \times A_2 \times ... \times A_n = \{(x_1, x_2, ..., x_n): x_1 \in A_1, x_2 \in A_2, ..., x_n \in A_n\}$$

In words: $A_1 \times A_2 \times ... \times A_n$ is the set of all ordered *n*-tuples in which the first element comes from A_1, the second from A_2, and so on.

---

[4] The name comes from Cartesius, the latinised version of the name of the French mathematician and philosopher René Descartes (1596–1650).

We can write the definition of $A_1 \times A_2 \times \ldots \times A_n$ more concisely by using predicate logic notation:

$$A_1 \times A_2 \times \ldots \times A_n = \{(x_1, x_2, \ldots, x_n): \forall\ i \in \{1, 2, \ldots, n\}\ (x_i \in A_i)\}$$

The notation here is a little different from the notation in Chapter 4, because the set of possible values of i is written down explicitly here.

EXAMPLE 5.3.2

Let $A = \{x, y\}$ and $B = \{1, 2, 3\}$. Write down the Cartesian product of A and B in enumerated form.

SOLUTION

$$A \times B = \{(x, 1), (x, 2), (x, 3), (y, 1), (y, 2), (y, 3)\}$$

Cartesian products arise in computing when we deal with strings of characters defined according to certain rules. For example, on some computers, the usercode that identifies a registered user must consist of 3 letters, followed by 3 digits, followed by a letter, for example XYZ123A. Let L denote the set of letters, and let D denote the set of digits; then the set of all valid usercodes is:

$$L \times L \times L \times D \times D \times D \times L$$

The usercode XYZ123A corresponds to the element (X, Y, Z, 1, 2, 3, A) of the set $L \times L \times L \times D \times D \times D \times L$. (The difference between writing XYZ123A and (X, Y, Z, 1, 2, 3, A) is just one of notation, not a conceptual difference.)

Sometimes we want to form a Cartesian product of a set with itself. This situation arises sufficiently often to warrant a notation of its own. If A is any set, the Cartesian product $A \times A \times \ldots \times A$ (n times) is written A^n. Thus we can write:

$$A^n = \{(x_1, x_2, \ldots, x_n): \forall\ i \in \{1, 2, \ldots, n\}\ (x_i \in A)\}$$

For example, $\mathbf{R}^2$ is the set of all ordered pairs of real numbers (x, y). In the study of co-ordinate geometry, this set is represented geometrically as a plane (the Cartesian plane), with x and y axes.

A second example, which we will use in Section 5.4, is the Cartesian product $\{0, 1\}^n$. The elements of this set are ordered n-tuples in which each element is either 0 or 1. For example, (1, 0, 0, 1, 0, 1, 1, 1) is an element of $\{0, 1\}^8$. We can think of $\{0, 1\}^n$ as the set of all strings of n bits.

5.4 Computer representation of sets

Some programming languages, such as Pascal, allow sets to be handled as a compound data type, where the elements of the sets belong to one of the data types available in the language, such as integers or characters. The question then arises: how are sets stored and manipulated in a computer?

A set is always defined in a program with reference to a universal set $\mathscr{E}$. We must make an exception here to the rule that the order of the elements of a set is irrelevant, because we will need to assume that the elements of $\mathscr{E}$ are listed in a definite order. Any set A arising in the program and defined with reference to this universal set $\mathscr{E}$ is a subset of $\mathscr{E}$. We want to know how the computer stores A internally.

The answer is that A is represented by a string of n bits, $b_1 b_2 ... b_n$, where n is the cardinality of $\mathscr{E}$. In the notation we have just introduced, the bit string $b_1 b_2 ... b_n$ can be regarded as the element $(b_1, b_2, ..., b_n)$ of $\{0, 1\}^n$. The bits are determined according to the rule:

$$b_i = 1 \text{ if the } i\text{th element of } \mathscr{E} \text{ is in } A$$

$$b_i = 0 \text{ if the } i\text{th element of } \mathscr{E} \text{ is not in } A$$

where i ranges over the values 1, 2, ..., n.

EXAMPLE 5.4.1

Let $\mathscr{E} = \{1, 2, 3, ..., 10\}$.

(a) Find the representation of $\{2, 3, 5, 7\}$ as a bit string.
(b) Find the set represented by the bit string 1001011011.

SOLUTION

(a) Looking in turn at each element of $\mathscr{E}$, and writing down 1 if the element is in $\{2, 3, 5, 7\}$ and 0 if it is not, we obtain the answer: 0110101000.
(b) The answer is obtained by writing down each element of $\mathscr{E}$ that corresponds to a 1 in the bit string: $\{1, 4, 6, 7, 9, 10\}$.

Each subset of an n-element universal set $\mathscr{E}$ can be paired up with the n-bit string that is its computer representation. By doing this, we set up a one-to-one correspondence between the subsets of $\mathscr{E}$ and all possible n-bit strings. We know already that there are 2^n such strings, so there must also be 2^n subsets of $\mathscr{E}$. This provides another proof that a set with n elements has 2^n subsets.

The operations of intersection, union and complement can be carried out directly on the bit strings, provided that the sets involved have been defined with reference to the same universal set. For example, the bit string of $A \cap B$ has a 1 wherever the bit strings of A and B both have a 1. This process for calculating the bit string of $A \cap B$ is called a *bitwise* **and** operation. Similarly, the bit strings of $A \cup B$ and $\overline{A}$ are calculated using a bitwise **or** and a bitwise **not** respectively.

EXAMPLE 5.4.2

Let the bit strings of A and B be 00101110 and 10100101 respectively. Find the bit strings of $A \cap B$, $A \cup B$ and $\overline{A}$.

SOLUTION

Performing the appropriate bitwise operations, we obtain the answers: 00100100, 10101111 and 11010001.

5.5 *Relations*

Relations between pairs of objects occur throughout mathematics and computing. We have already met some examples of relations. If one logical expression is equivalent to another, this is an example of a relationship that exists between those two logical expressions. We can refer to the *relation* of one logical expression being equivalent to another. In a similar way, we can refer to the relation of one set being a subset of another. Other examples in mathematics are the relation of one number being less than another, and the relation of one integer being divisible by another. In a database in which words are to be sorted into alphabetical order, we deal with the relation of one word preceding another in an alphabetical listing. A non-mathematical example is the various relationships between members of a family — one person may be the sister of another, the cousin of another, and so on.

In each of these examples, a statement is being made about a pair of objects that is true in some cases and false in others; the statement 'x is less than y' is true if $x = 3$ and $y = 4$, for example, but false if $x = 3$ and $y = 2$.

Before we can develop these ideas further, we need to make them more precise. We notice that each of the relations we have mentioned is associated with a set, and that the relation makes sense only when it refers to ordered pairs of elements of that set. For example, the relation 'less than' makes sense when it refers to ordered pairs of elements of **R**, the set of real numbers; if (x, y) is an ordered pair of real numbers, then the statement 'x is less than y' is either true or false. Similarly, the relation 'is a sister of' makes sense when applied to ordered pairs of elements of the set of all people, while the relation 'precedes in alphabetical order' refers to ordered pairs of words. In general, if the set to which the relation applies is denoted by A, then the ordered pairs are elements of the Cartesian product $A \times A$.

Informally, then, a relation on a set A is a statement about ordered pairs (x, y) belonging to $A \times A$. The statement must be either true or false for each particular pair of values of x and y. This is often the most convenient way of thinking about a relation. The formal mathematical definition is somewhat different, however.

Definition

A *binary*[5] *relation* on a set *A* is a subset of *A* × *A*.

The above definition needs some explanation, since the connection between it and the informal idea of a relation might not be obvious. The easiest way to do this is by means of an example. Consider the relation 'less than' on the set $A = \{1, 2, 3, 4\}$. For any ordered pair of elements of *A* (i.e. for any element of *A* × *A*), the relation 'less than' is either true or false for that pair. We can list all of the ordered pairs for which the relation is true:

$$(1, 2), (1, 3), (1, 4), (2, 3), (2, 4), (3, 4)$$

Formally, the relation is the set of these ordered pairs:

$$\{(1, 2), (1, 3), (1, 4), (2, 3), (2, 4), (3, 4)\}$$

This set is a subset of *A* × *A*, and so it is a relation according to the definition given above.

If we denote this set of ordered pairs by *R*, then we can state that *x* and *y* are related by writing $(x, y) \in R$. In practice, it is more usual to write $x R y$ to mean that *x* and *y* are related. In fact, we don't even need the name *R* in this example — the relation already has its own symbol, '<', so we can simply write $x < y$ to mean that *x* is related to *y*.

There is a way of depicting a relation graphically that is often useful, provided that the set on which the relation is defined is not too large. This is done by using dots to represent the elements of the set, and drawing an arrow from *x* to *y* for each pair (x, y) for which *x* is related to *y*. The resulting diagram is an object known as a *directed graph*.

EXAMPLE 5.5.1

Draw the graphical representation of the relation 'less than' on $\{1, 2, 3, 4\}$.

SOLUTION

This is shown in Figure 5.10.

Figure 5.10

---

[5]The word 'binary' refers to the fact that the relation is between *two* elements of *A*. As this is the only kind of relation we will be studying, we will omit the word 'binary' in what follows.

A finite relation can also be represented as a rectangular array of Ts and Fs, known as a *relation matrix* (plural: *matrices*). The rows and columns of the matrix are labelled with the elements of the underlying set. If x is related to y, then the entry in the row labelled x and the column labelled y is T, otherwise it is F. This representation is particularly useful if a relation is to be stored and manipulated in a computer.

EXAMPLE 5.5.2

| Write down the relation matrix for the relation in Example 5.5.1.

SOLUTION

$$
\begin{array}{c c}
& \begin{array}{cccc} 1 & 2 & 3 & 4 \end{array} \\
\begin{array}{c} 1 \\ 2 \\ 3 \\ 4 \end{array} &
\begin{bmatrix}
F & T & T & T \\
F & F & T & T \\
F & F & F & T \\
F & F & F & F
\end{bmatrix}
\end{array}
$$

We can define different types of relations according to their properties. The most important of these for our future work are listed below.

Definitions

Let R be a relation on a set A.
- R is *reflexive* if $x R x$ for all $x \in A$.
- R is *irreflexive* if there are no elements x of A for which $x R x$.
- R is *symmetric* if $x R y$ implies $y R x$, for all $x, y \in A$.
- R is *antisymmetric* if $x R y$ and $y R x$ imply $x = y$, for all $x, y \in A$.
- R is *transitive* if $x R y$ and $y R z$ imply $x R z$, for all $x, y, z \in A$.

The above definitions could all be written somewhat more concisely using the notation of predicate logic, at the expense of producing what might appear to some people as an 'alphabet soup' of mathematical symbols. For example, the last definition (transitivity) could be written:

$$\forall x \, \forall y \, \forall z \, \{[(x R y) \wedge (y R z)] \to (x R z)\}$$

This more concise form (or something similar) would be used in programming in a relational database, in logic programming, and in formal system specifications.

As a general rule, we will use whatever combination of English words and mathematical symbols seems to make the meaning clearest. In this case, the definitions are probably easier to understand in the form in which they are given.

EXAMPLE 5.5.3

The following relations are defined on the set of all people. Classify them according to the definitions given above:

(a) 'is a sister of'
(b) 'is the father of'
(c) 'has the same parents as'

SOLUTION

(a) This relation is not reflexive; in fact it is irreflexive, because no person can be her (or his!) own sister. It is not symmetric (if X is the sister of Y then Y need not be the sister of X — Y could be the brother of X), nor is it antisymmetric (X can be the sister of Y and Y the sister of X, without X and Y being the same person). It is not transitive either, because if X is the sister of Y and Y is the sister of Z then X need not be the sister of Z. (This is a bit subtle; can you see why it is true?)
(b) This relation is not reflexive, it is irreflexive, it is not symmetric, it is antisymmetric (in a vacuous sense — think about the truth table for **if-then**), it is not transitive.
(c) This relation is reflexive (and not irreflexive), symmetric (and not antisymmetric) and transitive.

EXAMPLE 5.5.4

Classify the following relations, which are defined on the set **J** of integers:

(a) 'is less than or equal to'
(b) 'is divisible by'
(c) 'has the same parity as' (i.e. both integers are odd or both are even)

SOLUTION

(a) $x \leq x$ is always true, so the relation is reflexive (and not irreflexive). If $x \leq y$ then it is never the case that $y \leq x$ except when $x = y$, so the relation is antisymmetric (and not symmetric). If $x \leq y$ and $y \leq z$ then $x \leq z$, so the relation is transitive.
(b) This relation is reflexive and transitive, but neither symmetric nor antisymmetric.
(c) This relation is reflexive, symmetric and transitive.

Look again at the last of the three relations in Examples 5.5.3 and 5.5.4. Both relations express an idea of 'sameness' — 'has the *same* parents as', 'is the *same* parity as'. Notice that these two relations are reflexive, symmetric and transitive.

Definition

A relation that is reflexive, symmetric and transitive is called an *equivalence relation*.

An equivalence relation is a relation that expresses the idea that two elements are the same in some sense. The relation 'is equivalent to' on the set of logical expressions is another example of an equivalence relation — two logical expressions are equivalent if they have the same truth table.

If an equivalence relation is defined on a set, then the elements that are related to each other can be grouped together into subsets. For example, if the relation is 'has the same parents as', defined on the set of all people, then all the people with one particular set of parents form one subset, the people with another set of parents form another subset, and so on. (Some of the subsets may contain only one person.) Each person belongs to exactly one subset. We say that the set of all people has been *partitioned* into disjoint subsets.

In the same way, the set of integers can be partitioned into subsets in which all the elements have the same parity. There will be two subsets: the set of even numbers and the set of odd numbers.

The process of forming a partition will not work if the relation is not an equivalence relation. For example, 'less than' is not an equivalence relation (because it is neither reflexive nor symmetric), and it makes no sense to define a 'set of numbers that are all less than each other'.

These ideas are summarised in the following definition and theorem.

Definition

Let *A* be a set. A *partition* of *A* is a set of subsets of *A* such that every element of *A* is an element of exactly one of the subsets.

Theorem

Let *A* be a set, and let *R* be an equivalence relation on *A*. For each element *x* of *A*, let *E(x)* be the subset of *A* defined by $E(x) = \{y \in A: y\,R\,x\}$. Then the set of all the subsets *E(x)* is a partition of *A*.

The subsets $E(x)$ are called the *equivalence classes* of the relation *R*.

Example 5.5.5 below illustrates these ideas. This example might seem rather abstract now, but we will refer to it again when we study modular arithmetic in Chapter 12. Modular arithmetic has important applications to topics such as information coding and the computer generation of pseudo-random numbers for simulation modelling.

EXAMPLE 5.5.5

Let R be the relation on the set $\mathbf{J}$ of integers defined by the rule: $x R y$ if $x - y$ is divisible by 4 (i.e. $x - y = 4n$ for some integer n). Show that R is an equivalence relation, and describe the equivalence classes.

SOLUTION

If x is any integer, then $x - x = 0$, which is divisible by 4. Therefore R is reflexive.

Suppose $x - y$ is divisible by 4. This means that there is an integer n such that $x - y = 4n$. Then $y - x = -4n$, which is divisible by 4. Therefore R is symmetric.

Suppose $x - y$ and $y - z$ are both divisible by 4. This means that there are integers m and n such that $x - y = 4m$ and $y - z = 4n$. Then $x - z = x - y + y - z = 4m + 4n = 4(m + n)$, which is divisible by 4. Therefore R is transitive.

Since R is reflexive, symmetric and transitive, R is an equivalence relation.

We begin the task of describing the equivalence classes by choosing an element of $\mathbf{J}$ and describing the equivalence class containing that element. Suppose we choose 0. The equivalence class $E(0)$ is $\{y \in \mathbf{J}: y R 0\}$; in words, it is the set of all the integers y such that $y - 0$ is divisible by 4. This is simply the set of all multiples of 4 (positive, negative and zero):

$$E(0) = \{..., -12, -8, -4, 0, 4, 8, 12, ...\}$$

Now we choose another element of $\mathbf{J}$, say 1. Then $E(1)$ will contain all the integers y such that $y - 1$ is divisible by 4:

$$E(1) = \{..., -11, -7, -3, 1, 5, 9, 13, ...\}$$

In the same way, we can find the equivalence class containing 2:

$$E(2) = \{..., -10, -6, -2, 2, 6, 10, 14, ...\}$$

Finally, we have the equivalence class containing 3:

$$E(3) = \{..., -9, -5, -1, 3, 7, 11, 15, ...\}$$

The process of finding the equivalence classes stops here, because the equivalence class containing 4 has already been found; it is the same as the one containing 0. Notice that the four equivalence classes form a partition of $\mathbf{J}$, because every integer can be found in exactly one equivalence class.

It is worthwhile studying carefully the process we have used here to show that R is an equivalence relation, because this approach is used in similar problems. In order to show that a relation is an equivalence relation, we need to do three things:

- show that it is reflexive
- show that it is symmetric
- show that it is transitive

The first step is to determine what that means *for the particular relation given in the problem*. This means that the definitions of 'reflexive', 'symmetric' and 'transitive' must be interpreted for the problem at hand. For example, in order to show that a relation R is transitive, we must show that if $x R y$ and $y R z$ then $x R z$. In Example 5.5.5, this meant we had to show that if $x - y$ is divisible by 4 and $y - z$ is divisible by 4 then $x - z$ is divisible by 4. Once we have established what needs to be shown, we are already halfway there; the process of actually showing it is not necessarily difficult.

Before we leave equivalence relations, we will look at a (fairly informal) proof of the theorem that the equivalence classes $E(x)$ form a partition of A. In order to carry out the proof, we need to show that every element of A belongs to exactly one equivalence class.

It is easy to show that any element x of A belongs to *at least* one equivalence class. Since R is reflexive, we have $x R x$, so $x \in E(x)$.

Showing that no element belongs to more than one equivalence class takes a bit more work. Suppose that $E(x)$ and $E(y)$ are two different equivalence classes, and suppose that they *do* have an element in common, say z. We now apply the following chain of reasoning:

Let w be any element of $E(x)$.
Then $w R x$ (by the definition of $E(x)$).
Since $z \in E(x)$, we also have $z R x$ (again using the definition of $E(x)$).
Therefore $x R z$ (because R is symmetric).
Since $w R x$ and $x R z$, we deduce that $w R z$ (because R is transitive).
Since $z \in E(y)$, we have $z R y$ (by the definition of $E(y)$).
Since $w R z$ and $z R y$, we deduce that $w R y$ (because R is transitive).
Therefore $w \in E(y)$ (by the definition of $E(y)$).

What we have just shown is that *any element of $E(x)$ must also be an element of $E(y)$*. A similar chain of reasoning shows that any element of $E(y)$ must be an element of $E(x)$ (just interchange the roles of x and y in the argument above). Therefore $E(x) = E(y)$, which contradicts the fact that $E(x)$ and $E(y)$ are two different equivalence classes. We conclude that the assumption that $E(x)$ and $E(y)$ have an element in common must be false. Therefore no element of A can belong to more than one equivalence class.

Another type of relation, called a partial order relation, also occurs in many situations in computing. Its definition is given below.

Definition

A relation is a *partial order* relation if it is reflexive, antisymmetric and transitive.

Here are some examples of partial order relations:

- the relation $\leq$ on the set of real numbers
- the relation $\subseteq$ on the power set of a set
- the relation 'is divisible by' on the set of natural numbers
- the relation 'is a subexpression of' on the set of logical expressions (with a given set of variables)

As an example of the last one, p, q, $p \wedge q$, $\neg p$ and $(p \wedge q) \vee \neg p$ are the subexpressions of $(p \wedge q) \vee \neg p$.

If a partial order relation is defined on a set, we can regard the elements of the set as forming a hierarchy in which some elements are 'bigger' in some sense while others are 'smaller'. The word 'partial' refers to the fact that not all pairs of elements need to be related one way or the other; for example, if A and B are sets, it is not necessarily true that either $A \subseteq B$ or $B \subseteq A$. By contrast, it *is* true that if x and y are real numbers, then either $x \leq y$ or $y \leq x$; we express this fact by saying that $\leq$ is a *total* order relation on **R**.

Partial order relations occur in many areas of computing. One example arises if we have a computer program consisting of a number of modules: the main program, the subprograms called by the main program, the subprograms called by these subprograms, and so on. We can define a relation R on the set of modules $\{M_1, M_2, ..., M_n\}$, using the rule: $M_i R M_j$ if M_i is in the calling sequence of M_j (i.e. M_j is the same module as M_i or M_j calls M_i or M_j calls a module that calls M_i or ...). You can check that R is reflexive and transitive. If R is not antisymmetric then circular calls are possible, such as two modules calling each other. The use of such 'recursive' calls must be avoided in programming languages that do not support recursion, but it can also be a powerful programming technique when it is available, as we will see when we study recursion in Chapter 7. If we are not using recursive calls, then R is a partial order relation.

Exercises

1. Write the following sets in enumerated form:
   (a) The set of all vowels.
   (b) $\{x \in \mathbf{N}: 10 \leq x \leq 20 \text{ and } x \text{ is divisible by } 3\}$
   (c) The set of all natural numbers that leave a remainder of 1 after division by 5.

2. Write the following sets in predicate form:
   (a) $\{4, 8, 12, 16, 20\}$

   (b) {000, 001, 010, 011, 100, 101, 110, 111}
   (c) {1, 4, 9, 16, 25, ...}

3. Let $A = \{1, \{1\}, \{2\}, 3\}$. Determine which of the following statements are true and which are false:

   (a) $1 \in A$ **(b)** $1 \subseteq A$
   (c) $\{1\} \in A$ **(d)** $\{1\} \subseteq A$
   (e) $\{\{1\}\} \subseteq A$ **(f)** $2 \in A$
   (g) $\{2\} \in A$ **(h)** $\{2\} \subseteq A$
   (i) $\{3\} \in A$ **(j)** $\{3\} \subseteq A$

4. Let $\mathscr{E} = \{x \in \mathbf{N}: x \leq 12\}$. Let $A = \{x: x \text{ is odd}\}$, $B = \{x: x > 7\}$, and $C = \{x: x \text{ is divisible by } 3\}$. Depict the sets on a Venn diagram. Hence write down the following sets in enumerated form:

   (a) $A \cap B$
   (b) $B \cup C$
   (c) $\overline{A}$
   (d) $(A \cup \overline{B}) \cap C$
   (e) $\overline{A \cup C \cup \overline{C}}$

5. Illustrate the first distributive law $A \cap (B \cup C) = (A \cap B) \cup (A \cap C)$ using Venn diagrams.

6. Illustrate the second absorption law $A \cup (A \cap B) = A$ using Venn diagrams.

7. Show that $\overline{A \cap B} = \overline{A} \cup \overline{B}$ using the laws of sets.

8. In a class of 35 students, there are 12 students who can speak German, and 5 students who can speak Japanese. If 2 of the students can speak both of these languages, how many of the students can speak neither language?

9. With the aid of a Venn diagram, obtain a version of the Principle of inclusion and exclusion for the cardinality of the union of three sets.

10. Is the statement 'a set with n elements has 2^n subsets' true when $n = 0$?

11. Let $A = \{a, b, c\}$ and $B = \{p, q\}$. Write down the following sets in enumerated form:

   (a) $A \times B$ **(b)** A^2 **(c)** B^3

12. Express each of the following sets as a Cartesian product of sets:

   (a) The set of all possible three-course meals (entrée, main course and dessert) at a restaurant.
   (b) The set of car registration plates consisting of three letters followed by three digits.
   (c) The set of all possible outcomes of an experiment in which a coin is tossed three times.

13. Let $\mathscr{E} = \{0, 1, 2, ..., 15\}$.

   (a) Find the representation of $\{2, 4, 5, 7, 11, 14\}$ as a bit string.
   (b) Write down the set represented by the bit string 1010 0110 1110 1001.
   (c) If A and B are represented by the bit strings 0011 0100 0110 1101 and 1010 1001 0001 0111, find the representations as bit strings of $A \cap B$, $A \cup B$, $\overline{A}$ and $\overline{B}$.

14. Write an algorithm to obtain the bit string representation of $A - B$ from the bit string representations of A and B, where A and B are subsets of a universal set $\mathscr{E}$ with n elements.

15. Let $A = \{1, 2, 3, 4, 5\}$, and let R be the relation on A defined as follows:

$$R = \{(1, 3), (1, 4), (2, 1), (2, 2), (2, 4), (3, 5), (5, 2), (5, 5)\}$$

    (a) Write down the matrix representation of R.
    (b) Draw the graphical representation of R.

16. Let R be the relation on $\{a, b, c, d\}$ defined by the following matrix:

$$\begin{array}{c} \\ a \\ b \\ c \\ d \end{array}\begin{array}{cccc} a & b & c & d \\ \begin{bmatrix} T & F & T & F \\ F & T & T & F \\ F & T & T & F \\ F & F & F & T \end{bmatrix} \end{array}$$

    (a) Draw the graphical representation of R.
    (b) State, giving reasons, whether R is reflexive, symmetric or transitive.

17. Is the matrix representation of a relation unique, or could the same relation be represented by two different matrices?

18. Determine whether each of the following relations is reflexive, irreflexive, symmetric, antisymmetric or transitive:

    (a) 'is a sibling (brother or sister) of', on the set of all people;
    (b) 'is the son of', on the set of all people;
    (c) 'is greater than', on the set of real numbers;
    (d) the relation R on the set of real numbers, defined by $x R y$ if $x^2 = y^2$;
    (e) 'has the same integer part as', on the set of real numbers;
    (f) 'is a multiple of', on the set of integers.

19. Determine which of the relations in Exercise 18 are equivalence relations. For those that are equivalence relations, describe the equivalence classes.

20. Determine which of the relations in Exercise 18 are partial order relations.

21. A computer program consists of five modules: $M_1, M_2, ..., M_5$. A relation R on the set of modules is defined by the rule: $M_i R M_j$ if M_i is in the calling sequence of M_j. The relation matrix for R is shown below:

$$\begin{array}{c} \\ M_1 \\ M_2 \\ M_3 \\ M_4 \\ M_5 \end{array}\begin{array}{ccccc} M_1 & M_2 & M_3 & M_4 & M_5 \\ \begin{bmatrix} T & F & T & T & F \\ F & T & T & F & F \\ F & F & T & F & F \\ F & F & T & T & F \\ F & F & T & T & T \end{bmatrix} \end{array}$$

    (a) Verify that R is reflexive, antisymmetric and transitive.
    (b) Which module is the main program?

6. FUNCTIONS

6.1 Functions and computing

If you have already studied functions in mathematics, it is likely that what you were studying was a type of function known as a 'real-valued function of a real variable'. If this is the case, then you probably think of a function as 'something that has a mathematical formula' such as $x^2 - 2x + 3$, or 'something you can draw the graph of' using x and y axes.

In computing (and in many areas of mathematics, for that matter), we take a different approach to functions, which turns out to be more useful. Our definition of a function is considerably more general, and in most cases it will not be possible to draw graphs of the functions we will be studying. The way we think about functions will also be somewhat different. The functions used in programming in high-level languages (both the built-in 'library' functions available in those languages and the function subprograms that you write yourself) are usually functions in this more general sense (with some qualifications, as we will see later). Functions of a real variable will be needed in Chapter 13 when we study the time complexity of algorithms.

Definition

> Let X and Y be sets. A *function* from X to Y is a rule that assigns to each element of X exactly one element of Y.

We will generally use lower case letters such as f, g and h to denote functions. Greek letters such as ϕ (phi) are also commonly used. If f is a function from X to Y, we indicate this by writing $f\colon X \to Y$. X is called the *domain* of f, and Y is the *codomain*.

If $f\colon X \to Y$ is a function with domain X and codomain Y, and if x is any element of X, then according to the definition there is exactly one element of Y assigned to x. That element is called the *image* of x, and is written $f(x)$.

For our first example, we will take a function you have probably seen before, and see how it can be viewed in the light of the definition.

Let the function f be defined as follows:

$$f\colon \mathbf{R} \to \mathbf{R}, \quad f(x) = x^2$$

This function assigns to each element of its domain $\mathbf{R}$ (i.e. to each real number) an element of its codomain (which is also $\mathbf{R}$). The rule that specifies which element to assign is $f(x) = x^2$, that is, the function f assigns to each real number the square of that number. If we pick any element of the domain, we can find the image of that element; for example, $f(5) = 25, f(-3) = 9, f(\sqrt{2}) = 2$. We see that every element of the domain **R** has a unique image, and each such image is an element of the codomain **R**. This ensures that f is a function, according to the definition.

A helpful way of thinking about this function is to imagine a machine that accepts input and produces output:

The function f is a 'squaring machine'. We may feed in any element of the domain (i.e. any real number) as input, and the machine will square it and output the result. We can think of the symbol f as standing for the *process* of squaring a number. The domain of f is the set of valid inputs, and the codomain is a set to which all of the possible outputs belong.

It is important to be clear about what the notation means — $f(x)$ denotes the element of the codomain that is the image of an element x of the domain, whereas f by itself is the name of the function, and represents the process that the function carries out. We could have written the rule for the function as $f(y) = y^2$ and it would still be the *same* function, because it carries out the same squaring process. In the terminology of predicate calculus, the variable x in the rule for the function is a *bound* variable, because the rule really means '$\forall x \, [f(x) = x^2]$'.

We want to make one more observation before we leave this example. It concerns the codomain **R**, to which all the images $f(x)$ belong. Notice that some elements of the codomain are not images of anything. For example, $-1 \in$ **R**, but -1 is not the image of any element of the domain, because the square of a real number cannot be negative. There is nothing in the definition of a function to say that all of the codomain has to be 'used'; the set of images of elements of the domain is a subset of the codomain, and it can be a proper subset.

If $f: X \rightarrow Y$ is any function, the set of images $\{y \in Y: y = f(x)$ for some $x \in X\}$ is called the *range* of f. The range of a function is a subset of the codomain.

In the example, the range of f is the set of non-negative real numbers: $\{y: y \geq 0\}$.

The function we have just been looking at is an example of a real-valued function of the kind you might study in a calculus course. In the next few examples we want to consider other types of functions, especially those that might occur in problems related to computing.

For our second example of a function, let $A = \{1, 2, 3\}$ and $B = \{1, 2, 3, 4\}$. Let f be the function defined in the following way:

$$f: A \rightarrow B, \quad f(1) = 3, \quad f(2) = 2, \quad f(3) = 2$$

This is a perfectly well defined function (although it is probably not a very useful one). It doesn't have a 'formula' like x^2, but that doesn't matter. All that matters is that *every element of the domain has exactly one image, and that image is an element of the codomain.* The elements of the domain A are 1, 2 and 3, their images are 3, 2 and 2 respectively, and these images are all in the codomain B. That is all that is needed in order for f to be a function. We see that the elements of B that occur as images are 2 and 3, so the range of f is $\{2, 3\}$.

While it would be possible to draw a graph of this function in the conventional sense, it would not be particularly useful to do so. A more appropriate way of depicting a function like this is by means of an arrow diagram. The arrow diagram for f is shown in Figure 6.1. Notice that exactly one arrow emerges from each element of the domain. This must always be the case if the diagram is to represent a function.

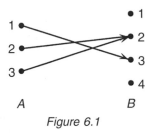

Figure 6.1

For our third example, let C be the set of all ASCII characters. Recall that each ASCII character has a unique character code in the range of integers from 0 to 127, and conversely that each integer in that range is the ASCII code of exactly one character. We can define the following functions:

$$ord: C \to \{0, 1, 2, \ldots, 127\}, \; ord(c) = \text{ASCII code of } c$$

$$chr: \{0, 1, 2, \ldots, 127\} \to C, \; chr(n) = \text{character with ASCII code } n$$

(The notation *ord* and *chr* for these functions is used in Pascal, and also in some spreadsheet languages such as Lotus 1-2-3.)

In order to find the image of an element of the domain of either of these functions, we would need to refer to a table of ASCII characters. For example, by looking up an ASCII table we find that $ord(\text{'A'}) = 65$, $ord(\text{'a'}) = 97$, and $ord(\text{'*'}) = 42$. It follows immediately that $chr(65) = \text{'A'}$, $chr(97) = \text{'a'}$, and $chr(42) = \text{'*'}$.

Each of the functions *ord* and *chr* has for its range the entire codomain of the function.

EXAMPLE 6.1.1

Let $X = \{\text{finite non-empty strings of bits}\}$ and $Y = \{0, 1, 2, 3, \ldots\}$. Determine whether the following functions are well defined:

(a) $f: X \to Y$, $f(s) = $ number of ones in s

(b) $g: X \rightarrow Y$, $g(s) =$ first bit of s
(c) $h: X \rightarrow Y$, $h(s) =$ position in the string of the leftmost zero of s
(d) $j: X \rightarrow X$, $j(s) =$ string obtained by appending 0 or 1 to s
(e) $k: Y \rightarrow X$, $k(n) =$ string of n ones

SOLUTION

(a) Given any finite non-empty string of bits, s, it is always possible to find the number of ones in s, and the result is always an element of Y. Therefore f is a function.
(b) The first bit of a finite non-empty string can always be found, and it is either 0 or 1. Since 0 and 1 are both elements of Y, g is a function.
(c) Some bit strings do not contain zeros, so h is not well defined.
(d) The image of an element is not uniquely determined, because either 0 or 1 can be appended, so j is not well defined.
(e) Given any element n of Y, it is possible to write down a string of n ones. However, if $n = 0$ the string will be the empty string, which is not an element of X. Therefore k is not well defined.

We have already seen examples in which the range of the function is not all of the codomain. The next definition refers to the situation where the range *is* the entire codomain.

Definition

A function is *onto* if its range is equal to its codomain.

Equivalently, a function is onto if every element of the codomain is the image of at least one element of the domain. (The use of 'onto' as an adjective may seem strange, but the terminology is well established.)

It is possible for two different elements of the domain of a function to have the same image in the codomain. In our first example (the 'squaring' function), 2 and –2 have the same image: $f(2) = 4$ and $f(-2) = 4$. The next definition provides the terminology for the functions for which this does not happen.

Definition

A function is *one-to-one* if no two distinct elements of the domain have the same image.

In order to show that a function f is one-to-one, we need to show that if x_1

and x_2 are elements of the domain and $x_1 \neq x_2$, then $f(x_1) \neq f(x_2)$. In order to show that a function is not one-to-one, on the other hand, it is sufficient to find two elements of the domain with the same image (as we did with the 'squaring' function).

EXAMPLE 6.1.2

For each of the following functions, determine whether the function is onto and whether it is one-to-one:

(a) $f: \mathbf{R} \to \mathbf{R}$, $f(x) = 2x + 1$
(b) the function *ord* defined earlier
(c) the function *chr* defined earlier
(d) the function f in Example 6.1.1(a)
(e) the function g in Example 6.1.1(b)
(f) $\phi: X \to X$, $\phi(s) = $ string obtained by appending 0 to s (with X as defined in Example 6.1.1)

SOLUTION

(a) If $y \in \mathbf{R}$, then we can set $y = 2x + 1$ and solve for x, obtaining $x = \dfrac{y-1}{2}$.

Therefore every element y of the codomain is the image of the element $\dfrac{y-1}{2}$ of the domain. Thus f is onto.

Suppose that there are two elements x_1 and x_2 with the same image: $f(x_1) = f(x_2)$. Then:

$$2x_1 + 1 = 2x_2 + 1$$

Hence $\qquad 2x_1 = 2x_2$

so $\qquad x_1 = x_2$

This is contrary to the fact that x_1 and x_2 are different elements. We conclude that it is impossible for two distinct elements to have the same image.[1] Therefore f is one-to-one.

(b) Every integer from 0 to 127 is the ASCII code of a character in the ASCII character set, so *ord* is onto. No two characters have the same code, so *ord* is one-to-one.

(c) Every character has an ASCII code, so *chr* is onto. No two codes correspond to the same character, so *chr* is one-to-one.

---

[1] What we have actually done here is to prove the statement: 'If $f(x_1) = f(x_2)$ then $x_1 = x_2$'. This is the contrapositive of 'If $x_1 \neq x_2$ then $f(x_1) \neq f(x_2)$', which is the definition of 'one-to-one', and we know from our work in logic that any **if-then** statement and its contrapositive are logically equivalent.

(d) Recall that the function is $f: X \to Y$, $f(s) =$ number of ones in s.

Is every element of Y the image of something? In other words, if we count the number of ones in a string of bits, could we get any of the numbers 0, 1, 2, ... as the result? The answer is yes, so f is onto.

Is it possible for two different elements to have the same image? In other words, could two *different* strings of bits yield the *same* result when we count the number of ones? The answer is clearly yes (101 and 110 is one of many examples), so f is not one-to-one.

(e) Recall that the function is $g: X \to Y$, $g(s) =$ first bit of s.

Is every element of Y the image of something? No; 2 is not the image of any element of X, so g is not onto.

Can two different elements have the same image? Yes; 00 and 01, for example, so g is not one-to-one.

(f) Is every element of X the image of something? No; 01 is not, because it doesn't end in 0. Therefore ϕ is not onto.

Can two different elements have the same image? If so, this would mean that two different bit strings would give the same bit string when 0 is appended to them, which is clearly impossible. Therefore ϕ is one-to-one.

The solution to Example 6.1.2 reinforces an important point made in Chapter 5 when we studied equivalence relations. You have already made a substantial step towards solving a mathematical problem when you have identified just what it is that you need to show. In order to determine whether a function is onto, you need to ask yourself: *Is every element of the codomain the image of something in the domain?* In order to determine whether a function is one-to-one, the question you need to ask is: *Can two different elements of the domain have the same image?* You then need to interpret the question in the context of the particular problem. This is what we were doing in Example 6.1.2.

Sketching an arrow diagram can also help you to decide whether a function is onto and whether it is one-to-one. Some examples of typical arrow diagrams for the various types of functions are shown in Figure 6.2.

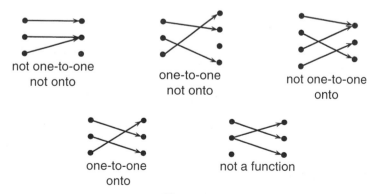

Figure 6.2

The last diagram in Figure 6.2 is not the arrow diagram of a function, because it fails the requirement that every element of the domain must have exactly one image. (Note that the terms 'onto' and 'one-to-one' apply only to functions. If you have determined that what you have been given is not a function, then the question of whether it is onto or one-to-one does not arise.)

6.2 *Composite functions and the inverse of a function*

Suppose *f* and *g* are two functions. If we think of *f* and *g* as machines with input and output, we could imagine linking them together so that the output of *f* becomes the input of *g*:

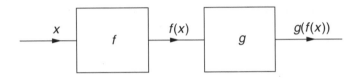

This will work only if the output from *f* belongs to the domain of *g*. In order to ensure that this is always the case, we will assume that the codomain of *f* equals the domain of *g*. Specifically, let *A*, *B* and *C* be arbitrary sets, and let *f*: *A* → *B* and *g*: *B* → *C*.

We can now think of the combination of the two machines as a single machine with input $x \in A$ and output $g(f(x)) \in C$. This new machine corresponds to a function from *A* to *C*, called the *composite* function of *f* and *g*.

The formal definition follows.

Definition

Let *f*: *A* → *B* and *g*: *B* → *C* be functions. The *composite function* of *f* and *g* is the function:

$$g \circ f: A \to C, (g \circ f)(x) = g(f(x))$$

Notice that *g* ∘ *f* needs to be read from right to left: it means *first* apply *f*, then apply *g* to the result.

Although real-valued functions are only of minor importance for our purposes, we will use them in our first example because you are likely to be more familiar with them.

EXAMPLE 6.2.1

Let $f: \mathbf{R} \to \mathbf{R}$, $f(x) = x^2$, and $g: \mathbf{R} \to \mathbf{R}$, $g(x) = 3x - 1$. Find $f \circ g$ and $g \circ f$.

SOLUTION

Note firstly that the composite function $f \circ g$ exists because the codomain of g equals the domain of f. Similarly, $g \circ f$ exists because the codomain of f equals the domain of g.

The function $f \circ g$ is found as follows:

$$f \circ g: \mathbf{R} \to \mathbf{R}, \; (f \circ g)(x) = f(g(x))$$
$$= f(3x - 1)$$
$$= (3x - 1)^2$$

where the last line is obtained by substituting $3x - 1$ in place of x in the formula for $f(x)$.

The function $g \circ f$ is obtained in a similar manner:

$$g \circ f: \mathbf{R} \to \mathbf{R}, \; (g \circ f)(x) = g(f(x))$$
$$= g(x^2)$$
$$= 3x^2 - 1$$

The two composite functions in the last example can be depicted in the following way:

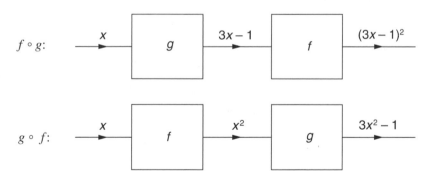

EXAMPLE 6.2.2

Let X be the set of all finite non-empty strings of characters. Let the functions f and g be defined as follows:

$$f: X \to \mathbf{N}, \; f(s) = \text{number of characters in } s$$

$$g: X \to X, \; g(s) = \text{string obtained by appending 'a' to } s$$

State whether the following functions exist. For those that do exist, describe the function as simply as possible.

(a) $f \circ f$
(b) $f \circ g$
(c) $g \circ f$
(d) $g \circ g$

SOLUTION

(a) The codomain of f does not equal the domain of f, so $f \circ f$ does not exist.

(b) The codomain of g equals the domain of f, so $f \circ g$ exists. The composite function appends 'a' to a string and counts the number of characters in the resulting string. It can be described in the following way:

$$f \circ g: X \rightarrow \mathbf{N}, \ (f \circ g)(s) = (\text{number of characters in } s) + 1$$

(c) The codomain of f does not equal the domain of g, so $g \circ f$ does not exist.

(d) The codomain of g equals the domain of g, so $g \circ g$ exists. The composite function appends 'a' to a string, then appends 'a' to the resulting string:

$$g \circ g: X \rightarrow X, \ (g \circ g)(s) = \text{string obtained by appending 'aa' to } s$$

We now turn to the problem of defining the inverse of a function. In Section 6.1 of this chapter, we defined the following two functions:

$$ord: C \rightarrow \{0, 1, 2, \ldots, 127\}, \ ord(c) = \text{ASCII code of } c$$

$$chr: \{0, 1, 2, \ldots, 127\} \rightarrow C, \ chr(n) = \text{character with ASCII code } n$$

There is a close relationship between *ord* and *chr*. Firstly, the domain of *ord* equals the codomain of *chr*, while the codomain of *ord* equals the domain of *chr*. Secondly, each function 'undoes' or reverses the effect of the other; *ord* converts a character to its corresponding code, while *chr* converts a code to its corresponding character. We express this fact by saying that *ord* and *chr* are *inverses* of each other.

This idea can be made more precise by looking at the composite functions *ord* ∘ *chr* and *chr* ∘ *ord*. The function *ord* ∘ *chr*: $\{0, 1, 2, \ldots, 127\} \rightarrow \{0, 1, 2, \ldots, 127\}$ converts a number to the corresponding character, and then converts the resulting character back to the original number. The overall effect of *ord* ∘ *chr* is to leave every number unchanged. Similarly, the function *chr* ∘ *ord*: $C \rightarrow C$ has the overall effect of leaving every character unchanged.

Before we can define the inverse of a function formally, we need another definition.

Definition

> Let *A* be a set. The *identity function* on *A* is the function:
>
> $$i: A \rightarrow A, \ i(x) = x$$

The identity function is a 'do nothing' function; it simply maps each element of *A* to the element itself. It can be thought of as a machine that outputs anything it receives as input.

While it has to be admitted that the identity function on a set is not a very interesting function, it does have some important properties. Firstly, it is one-to-one and onto. Secondly, if *f* is any function with domain *A*, and if *i* denotes the identity function on *A*, then *f* ∘ *i* is the same function as *f* itself. Similarly, if *g* is a function with codomain *A*, then *i* ∘ *g* is the same function as *g*.

We can now give the formal definition of the inverse of a function.

Definition

> Let *f*: *A* → *B* and *g*: *B* → *A* be functions. If *g* ∘ *f*: *A* → *A* is the identity function on *A*, and if *f* ∘ *g*: *B* → *B* is the identity function on *B*, then *f* is the *inverse* of *g* (and *g* is the inverse of *f*).

Not every function has an inverse, as we will see shortly. If a function does have an inverse, it can have only one.

The inverse of a function *f* is denoted by f^{-1}. It is best to think of this simply as the notation for the inverse of a function as defined above. Don't think of it as '*f* to the power of –1', as it is quite different from raising a *number* to the power –1, for example $2^{-1} = \frac{1}{2}$. In particular, we never write $\frac{1}{f}$ for the inverse of *f*.

The domain of f^{-1} is the codomain of *f*, and vice versa. We can think of f^{-1} as a machine that 'reverses' *f* — it takes any valid output of *f* as its own input, and produces as output the corresponding input of *f*.

EXAMPLE 6.2.3

| Let *f*: {a, b, c} → {1, 2, 3}, *f*(a) = 2, *f*(b) = 3, *f*(c) = 1. Find f^{-1}.

SOLUTION

The domain of f^{-1} is the codomain of f and vice versa. The rule for f^{-1} is obtained by reversing the rule for f:

$$f^{-1}: \{1, 2, 3\} \rightarrow \{a, b, c\}, \ f^{-1}(1) = c, f^{-1}(2) = a, f^{-1}(3) = b$$

The arrow diagrams for the functions f and f^{-1} in Example 6.2.3 are shown in Figure 6.3. The diagram for the inverse is obtained by reversing the arrows; the arrow from c to 1 becomes an arrow from 1 to c, and so on.

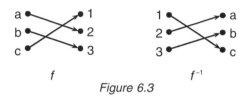

Figure 6.3

The arrow diagram gives us a clue to the situations in which a function has no inverse. If f is not onto, the arrow diagram obtained by reversing the arrows will have elements of the domain with no image, and therefore it cannot represent a function. If f is not one-to-one, the reversed diagram will have at least one element with two images, so it cannot represent a function in this case either. This observation is summarised in the following theorem.

Theorem

A function f has an inverse if and only if f is onto and one-to-one.

If a function $f: X \rightarrow Y$ is onto and one-to-one, then f establishes an exact one-to-one correspondence between the elements of X and the elements of Y. It follows that X and Y must have the same cardinality. We will see a practical consequence of this fact shortly.

EXAMPLE 6.2.4

Determine which of the following functions have inverses:
(a) $f: \mathbf{R} \rightarrow \mathbf{R}, \ f(x) = 2x + 1$
(b) $g: \mathbf{R} \rightarrow \mathbf{R}, \ g(x) = x^2$
(c) $h: \{x \in \mathbf{R}: x \geq 0\} \rightarrow \{x \in \mathbf{R}: x \geq 0\}, \ h(x) = x^2$

SOLUTION

(a) This function is onto and one-to-one, so f^{-1} exists. To find a formula for the inverse function, write $x = 2f^{-1}(x) + 1$ and solve for $f^{-1}(x)$ to obtain

$f^{-1}(x) = \dfrac{x-1}{2}$. Intuitively, f 'doubles and adds 1', while f^{-1} 'subtracts 1 and divides by 2', which is the reverse process.

As a check on the answer, note that $(f \circ f^{-1})(x) = f(f^{-1}(x)) = f\left(\dfrac{x-1}{2}\right) = 2\left(\dfrac{x-1}{2}\right) + 1 = x$, and similarly that $(f^{-1} \circ f)(x) = x$.

(b) This function is neither onto nor one-to-one, so it has no inverse.

(c) Notice that this is not the same function as the one in (b), because both the domains and the codomains are different. The function h is one-to-one, because it is impossible for two non-negative real numbers to have the same square. The function is also onto. The rule for the inverse function is $h^{-1}(x) = \sqrt{x}$.

Here is an example of a practical situation where inverse functions arise. It is often desirable to encrypt a confidential message prior to transmitting it via a possibly insecure channel, in order to ensure that it cannot be read by an unauthorised person. Let X be the set of all possible original messages, and let Y be the set of all encrypted messages. Then the code used to carry out the encryption process can be thought of as a function $f: X \to Y$. In order to ensure that any message can be decoded, there must be an inverse function $f^{-1}: Y \to X$ to carry out the decoding process. This means that any function f that we might consider using to perform the encryption must be one-to-one and onto.

A similar situation arises with software for file compression. The process of compressing a file so that it occupies less space on a disk can be thought of as an encryption process corresponding to a function f. The data in a file is stored on a disk as a finite (and non-empty) string of bits, and we will assume that there is an upper limit N on the number of bits in any file. If we also assume that no string of N bits or fewer can be ruled out as the possible contents of a file, then the domain of f is the set X of all bit strings with N bits or less. The codomain is the set Y of those bit strings that can occur as the contents of a compressed file.

The function f must have an inverse function f^{-1} to perform the process of expanding a compressed file to retrieve the original data. This means that f must be one-to-one and onto, and therefore that X and Y must have the same cardinality. Of course, we would also like the function f to have the property that each compressed file is smaller than the corresponding original file. Can we find such a function?

The answer is no! If each compressed file is smaller than the original one, then Y must contain only files with fewer than N bits. This makes Y a proper subset of X, so Y must contain fewer elements than X. We arrive at the following surprising conclusion: *there is no file compression algorithm that compresses every file*.

It is actually possible to prove an even stronger statement: any file compression algorithm that makes at least one file smaller must also make at least one file larger.

The file compression utilities in common use can be very effective in compressing most of the files that occur in practice. With any such program, however, there must always be files that cannot be compressed with that program.

6.3 *Functions in programming languages*

At the beginning of this chapter, we stated that many programming languages (including spreadsheet and database languages) include a construct called a 'function' as part of the language. There are usually a number of 'built-in' functions available in a function 'library', and there is often the means available for programmers to write their own functions. How are these functions related to functions in the mathematical sense?

For example, the spreadsheet software Lotus 1-2-3 includes a comprehensive library of functions for manipulating data stored in a spreadsheet. Functions in Lotus 1-2-3 are identified by a name beginning with the character '@', such as @ABS, @ISSTRING and @LEFT. These Lotus functions can be interpreted as functions in the sense we have been using in this chapter.

The function @ABS takes a real number x as input, and returns its absolute value $|x|$ as output. We can therefore take the domain and codomain to be the set of real numbers, and write:

$$@\text{ABS}: \mathbf{R} \to \mathbf{R}$$

Strictly speaking, we are using $\mathbf{R}$ here to denote the set of *computer representations* of real numbers, rather than the set of real numbers in the mathematical sense.

The function @ISSTRING is a little more complicated. It can take as input an item of data of any type (character string, number value or logical condition), or (more usually) the address of a cell in the spreadsheet. The output is 1 if the input data (or the data contained in the cell address) is a string, and 0 if it is not. If X denotes the set of all valid items of data and all valid cell addresses, then:

$$@\text{ISSTRING}: X \to \{0, 1\}$$

The function @LEFT is described in Lotus 1-2-3 terminology as a 'function with two arguments'. The input to @LEFT consists of a character string s and a natural number n, and the output is the string consisting of the first n characters of s; for example, @LEFT("Hello",3) = "Hel". If we want to interpret @LEFT as a function in the mathematical sense, we must write the domain as a Cartesian product. If S denotes the set of all character strings, then:

$$@\text{LEFT}: S \times \mathbf{N} \to S$$

The term 'function' in Lotus 1-2-3 also embraces constants such as @PI, which returns a numerical approximation to π. This 'function' has no arguments (think of a machine that outputs the value of π without needing any input). Somewhat artificially, we can think of @PI as a function whose domain contains one (unspecified) element:

$$\texttt{@PI}: \{*\} \rightarrow \mathbf{R}, \texttt{@PI}(*) = 3.1415926$$

This is really stretching our definition of 'function'!

The programming language Pascal, in addition to having a library of built-in functions, also allows you as the programmer to write your own functions. The syntax rules for the language require you to specify the data types of the input and output in the statement that heads the function subprogram. A typical function header in Pascal might look like this:

```
function f(x: char; y: real): integer;
```

The body of the function follows the header, and includes an algorithm for evaluating the function in terms of the arguments x and y.

If we treat this as a mathematical function f, then the domain of f is a subset of $C \times \mathbf{R}$, where C denotes the set of all characters available on the machine, and we can take the codomain to be $\mathbf{J}$.

In Pascal and other programming languages, however, it is possible for the value returned by a function to depend on more than just the arguments (it can depend on the value of a 'global' variable, for example). A function can also produce 'output' in addition to the value it returns (by changing the value of a global variable, for example, or by printing a message). Because of these 'side-effects', a function can return two different results when it is called twice with the same arguments. For this reason, a function in a programming language may not be a function in the mathematical sense.

One example of a function with side-effects is a pseudo-random number generator. Pseudo-random number generators are available as standard library functions in many programming languages, and are widely used in simulation software. A typical pseudo-random number generator might produce a different real number in the interval from a to b each time it is called with arguments a and b. At first sight, it might appear to be a function in the mathematical sense:

random: $\mathbf{R} \times \mathbf{R} \rightarrow \mathbf{R}$, *random*$(a, b)$ = a random number between a and b

When we use *random* in a program, we might find, for example, that *random*$(0, 1) = 0.3982473$ on the first call and *random*$(0, 1) = 0.8194702$ on the second call. Since the result is not uniquely determined by the values of the two arguments, we conclude that *random* does not qualify as a function in the mathematical sense of the word.

In this chapter, we have introduced the powerful notion of a function from any set to any other set, and we have seen how functions in this general sense arise in computing contexts. Functions play a central role in mathematics, and they will appear frequently in the chapters that follow.

> **Exercises**

1. Determine whether the following functions are well defined. For those that are well defined, state the domain, codomain and range.

   (a) $f: \mathbf{R} \rightarrow \mathbf{R}$, $f(x) = \dfrac{x}{3}$

   (b) $g: \mathbf{J} \rightarrow \mathbf{J}$, $g(x) = \dfrac{x}{3}$

   (c) $h: \{1, 2, 3\} \rightarrow \{1, 2, 3\}$, $h(n) = n + 1$

   (d) $d: \mathbf{N} \rightarrow \mathbf{N}$, $d(n) =$ number of digits in the decimal representation of n

   (e) $\phi: X \rightarrow \mathbf{R}$, $\phi(x) = \dfrac{1}{x}$, where X is the 'maximal' domain (i.e. the largest subset of $\mathbf{R}$ such that $\dfrac{1}{x}$ is well defined when $x \in X$)

   (f) $ispositive: \mathbf{R} \rightarrow \{T, F\}$, $ispositive(x) = \begin{cases} T \text{ if } x \text{ is positive} \\ F \text{ if } x \text{ is negative} \end{cases}$

   (g) $r: \mathbf{N} \rightarrow \mathbf{J}$, $r(n) =$ remainder after n is divided by 6

   (h) $\psi: S \rightarrow S$, $\psi(s) =$ string obtained by removing the last character from s (where S is the set of finite non-null character strings)

2. Determine which of the following functions are one-to-one and which are onto:

   (a) $f: S \rightarrow S$, $f(s) =$ string obtained by reversing the order of the characters of s ($S = \{$finite non-null character strings$\}$)

   (b) $g: \mathbf{R} \times \mathbf{R} \rightarrow \mathbf{R}$, $g((x, y)) = x + y$

   (c) $s: \mathbf{N} \rightarrow \mathbf{N}$, $s(n) = n + 1$

   (d) $h: \{$English words$\} \rightarrow \{$letters$\}$, $h(w) =$ first letter of word w

   (e) $b: \mathbf{N} \rightarrow \{$finite non-null bit strings$\}$, $b(n) =$ binary representation of n (written without leading zeros)

   (f) $card: \mathcal{P}(A) \rightarrow \mathbf{N} \cup \{0\}$, $card(X) =$ the cardinality of the set X (where A denotes any finite set)

3. A function $f: \{1, 2, 3, 4, 5\} \rightarrow \{0, 1, 2, 3, 4\}$ is defined by the rule: $f(n)$ is the remainder after $3n$ is divided by 5. Draw an arrow diagram for this function. Hence state whether f is one-to-one and whether f is onto.

4. For the purpose of error detection, numeric codes (such as ID numbers) often include a final 'check digit'.

   Suppose a numeric code consists of a string of 9 digits $x_1 x_2 \ldots x_9$, followed by a final check digit x_{10} defined to be the rightmost decimal digit of $x_1 + 2x_2 + 3x_3 + \ldots + 9x_9$.

   (a) Verify that 2516238674 is a valid code.

   (b) Let X be the set of all strings of 9 digits, let Y be the set of all digits, and let $f: X \rightarrow Y$ be the function that assigns the correct check digit to each string, for example $f(251623867) = 4$. State, giving reasons, whether f is one-to-one and whether f is onto.

(c) If an error is made in keying in a code, will the check digit always detect it? Explain, with reference to your answer to (b).

5. Let the functions f, g and h be defined as follows:

$$f: \mathbf{R} \to \mathbf{R}, \ f(x) = 4x - 3$$

$$g: \mathbf{R} \to \mathbf{R}, \ g(x) = x^2 + 1$$

$$h: \mathbf{R} \to \mathbf{R}, \ h(x) = \begin{cases} 1 \text{ if } x \geq 0 \\ 0 \text{ if } x < 0 \end{cases}$$

Find rules for the following functions with domain and codomain $\mathbf{R}$:

(a) $f \circ f$ (b) $f \circ g$
(c) $g \circ f$ (d) $f \circ h$
(e) $h \circ f$ (f) $g \circ h$
(g) $h \circ g$

6. Find the inverse function of each of the following functions, or explain why no inverse exists:

(a) $f: \mathbf{R} \to \mathbf{R}, \ f(x) = 3x + 2$
(b) $abs: \mathbf{R} \to \mathbf{R}, \ abs(x) = |x|$
(c) $g: \mathbf{N} \to \mathbf{N}, \ g(n) = \begin{cases} n + 1 \text{ if } n \text{ is odd} \\ n - 1 \text{ if } n \text{ is even} \end{cases}$
(d) $h: S \to S, \ h(s) =$ the string obtained by moving the last character to the beginning of the string, for example $h(\text{'abcd'}) = \text{'dabc'}$ (where S is the set of finite non-null character strings)

7. Let X be the set of all names of students in a database maintained by a university. (Assume that no two students have the same name.) Let Y be the set of ID numbers of the students. The functions f and g are defined as follows:

$$f: X \to Y, \ f(x) = \text{ID number of student with name } x$$

$$g: Y \to \mathbf{N}, \ f(y) = \text{age (in years) of student with ID number } y$$

(a) Describe the functions $g \circ f$ and f^{-1}.
(b) Explain why g^{-1} does not exist.

8. Let S be the set of all finite non-null strings of characters. Let $upr: S \to S$ be the function that converts all lower-case letters to upper-case (and leaves all other characters unchanged). Similarly, let $lwr: S \to S$ be the function that converts all upper-case letters to lower-case.

(a) Evaluate $upr(\text{'Barbara Hill'})$ and $(lwr \circ upr)(\text{'Barbara Hill'})$.
(b) Are upr and lwr inverses of each other? Explain.

9. Let $X = \{1, 2, 3, \ldots, 20\}$, and let Y be the set of non-null strings of up to 20 characters. Let f be defined as follows:

$$f: X \to Y, \ f(x) = \text{English word for } x \text{ (in lower-case letters)}$$

For example, $f(1) =$ 'one'.

(a) State, giving reasons, whether:
   (i) f is one-to-one;
   (ii) f is onto;
   (iii) f^{-1} exists.

(b) If $g: Y \rightarrow X$, $g(s) =$ number of characters in s, evaluate the following expressions, where possible:
   (i) $(g \circ f)(7)$
   (ii) $(g \circ f)($'seven'$)$
   (iii) $(f \circ g)(7)$
   (iv) $(f \circ g)($'seven'$)$

10. Let $f: \mathbf{N} \rightarrow \mathbf{N}$, $f(n) =$ digital root of n. (Digital roots were defined in Chapter 1, Exercise 8.)

(a) State, with reasons, whether f is one-to-one and whether f is onto.

(b) Does f^{-1} exist? If so, describe it. If not, give a reason.

(c) Does $f \circ f$ exist? If so, describe it. If not, give a reason.

7. INDUCTION AND RECURSION

7.1 *Recursion and sequences*

The main purpose of this chapter is to introduce recursion. Recursion is a simple yet powerful idea that can be enormously useful in developing algorithms for solving complex problems. This is especially true if an algorithm is to be implemented in Lisp or a related programming language such as Scheme. We will also introduce the method of proof by induction, a technique closely related to recursion. In addition to being one of the standard proof techniques in mathematics, induction is a useful technique for verifying the correctness of algorithms.

A convenient way to introduce recursion is by considering infinite sequences of numbers, and this is the approach we will take here.

A *sequence* is a list of numbers in a particular order. A general sequence can be written down in the following way:

$$t(1), t(2), t(3), t(4), \ldots$$

The numbers $t(1)$, $t(2)$, $t(3)$, and so on are called the *terms* of the sequence. The purpose of the three dots (ellipsis) at the end is to indicate that the sequence can be regarded as continuing indefinitely.

There is usually a definite pattern to the numbers in the kinds of sequences that arise in practice. For example, the positive even numbers form a sequence:

$$2, 4, 6, 8, 10, 12, \ldots$$

In this example, $t(1) = 2$, $t(2) = 4$, $t(3) = 6$, and so on.

A sequence can be defined by giving a formula for the general term $t(n)$ in terms of the variable n. For the sequence given above, the general term is given by the formula $t(n) = 2n$.

There is a good reason for using 'function-style' notation for the general term of a sequence. We can think of a sequence as a particular kind of function; for example, the sequence 2, 4, 6, 8, 10, 12, ... can be identified with the following function from the set **N** of natural numbers to the set **R** of real numbers:

$$t: \mathbf{N} \to \mathbf{R}, \ t(n) = 2n$$

For convenience, we will always use **R** for the codomain of a sequence, even though in the present example all the terms of the sequence are natural numbers.

We can say, then, that a sequence is a function with domain **N** and codomain **R**; in fact, this is the formal mathematical definition of a sequence.

Here are some more examples of sequences, together with formulae for their general terms:

- 3, 5, 7, 9, 11, 13, ... $t(n) = 2n + 1$
- 0.2, 0.9, 1.6, 2.3, 3.0, 3.7, ... $t(n) = 0.7n - 0.5$
- 1, 4, 9, 16, 25, 36, ... $t(n) = n^2$
- 0, 1, 0, 1, 0, 1, ... $t(n) = \begin{cases} 0 & \text{if } n \text{ is odd} \\ 1 & \text{if } n \text{ is even} \end{cases}$

We have defined each of these sequences by stating a rule for the *n*th term $t(n)$ as a mathematical formula containing *n*. Definitions of this kind are called *non-recursive*, in order to distinguish them from recursive definitions, which we will introduce shortly.

A non-recursive definition can form the basis of an algorithm for generating the first *m* terms of a sequence. For example, the algorithm below generates the first *m* terms of the sequence 2, 4, 6, 8, 10, 12, ...:

1. Input *m*
2. **For** $n = 1$ **to** *m* **do**
   2.1. $t \leftarrow 2n$
   2.2. Output *t*

This is not the only way that the sequence 2, 4, 6, 8, 10, 12, ... can be generated, however. Here is another algorithm for generating the first *m* terms of the same sequence:

1. Input *m*
2. $t \leftarrow 2$
3. Output *t*
4. **For** $n = 2$ **to** *m* **do**
   4.1. $t \leftarrow t + 2$
   4.2. Output *t*

This algorithm is based on another way of describing the sequence 2, 4, 6, 8, 10, 12, ...; *the first term is 2, and each subsequent term is obtained by adding 2 to the previous term.* We can write this in mathematical notation in the following way:

$$t(1) = 2$$

$$t(n) = t(n - 1) + 2 \quad (n > 1)$$

These two equations together form what is called a *recursive* definition; the definition is recursive because the formula for $t(n)$ contains the previous term $t(n - 1)$. The equation $t(1) = 2$ forms a *base* or starting point for the recursion.

Notice that we are not permitted to substitute $n = 1$ into the recursive part of the definition, $t(n) = t(n - 1) + 2$; in fact, it would not make sense to do so, because there is no term $t(0)$.

It is important to understand that the recursive and non-recursive definitions are merely two different ways of defining the same sequence. Either definition can be used to find any particular term of the sequence, and both definitions have their advantages, depending on the circumstances.

For example, we can evaluate $t(4)$ using the recursive definition:

$$\begin{aligned}
t(4) &= t(3) + 2 && \text{(putting } n = 4 \text{ in the formula for } t(n)\text{)}\\
&= t(2) + 2 + 2 && \text{(putting } n = 3\text{)}\\
&= t(1) + 2 + 2 + 2 && \text{(putting } n = 2\text{)}\\
&= 2 + 2 + 2 + 2 && \text{(because } t(1) = 2\text{)}\\
&= 8
\end{aligned}$$

Alternatively, we can evaluate $t(4)$ using the non-recursive definition:

$$t(4) = 2 \times 4 = 8$$

EXAMPLE 7.1.1

Find a recursive and a non-recursive definition for the following sequence:

$$2, 5, 8, 11, 14, 17, \ldots$$

SOLUTION

A problem like this only makes sense if the terms form a clearly recognisable pattern. The rule for this sequence is clear — begin with 2, then go up in steps of 3. This immediately yields the recursive definition. Denoting the nth term by $t(n)$ as usual, we have:

$$t(1) = 2$$

$$t(n) = t(n - 1) + 3 \quad (n > 1)$$

In order to find a non-recursive definition, we begin with the observation that the terms of the sequence go up in steps of 3. This suggests that the formula for $t(n)$ should contain $3n$. However, $t(n) = 3n$ doesn't quite do the job; $t(n) = 3n$ is the rule for the sequence 3, 6, 9, 12, ... In order to obtain a sequence that starts at 2 rather than 3, we need to subtract 1 from each of these terms. The non-recursive definition is therefore:

$$t(n) = 3n - 1$$

The method we have shown here for finding the non-recursive definition is somewhat *ad hoc*, and in a more complicated problem it might be difficult to be certain that we have obtained the right formula. We will see later how the technique of proof by induction can be used to show that a non-recursive formula that we suspect to be true is in fact correct.

EXAMPLE 7.1.2

Evaluate $t(2)$, $t(3)$ and $t(4)$ for the following recursively defined sequence:

$$t(1) = 3$$
$$t(n) = 2t(n-1) + n \quad (n > 1)$$

SOLUTION

To evaluate $t(2)$:

$$t(2) = 2t(1) + 2 = 2 \times 3 + 2 = 8$$

To evaluate $t(3)$:

$$t(3) = 2t(2) + 3 = 2 \times 8 + 3 = 19$$

(Notice that we use the value of $t(2)$ we have already obtained.)
To evaluate $t(4)$:

$$t(4) = 2t(3) + 4 = 2 \times 19 + 4 = 42$$

EXAMPLE 7.1.3

Write an algorithm to output the first m terms of the sequence in Example 7.1.2.

SOLUTION

1. Input m
2. $t \leftarrow 3$
3. Output t
4. **For** $n = 2$ **to** m **do**
   4.1. $t \leftarrow 2t + n$
   4.2. Output t

There is a small but important matter of terminology that needs to be made clear. Although the algorithm in Example 7.1.3 is based on a recursive definition of the sequence, it is not a recursive algorithm but an *iterative* algorithm. ('Iterate' means to apply the same process repeatedly, as in a **For-do** loop.) We will look at recursive algorithms later in this chapter.

In all the examples of recursive definitions we have seen so far, the expression for $t(n)$ contains only the immediately preceding term $t(n-1)$. While this is the most usual situation in practice, in general a recursive formula for $t(n)$ can contain *any* of the earlier terms of the sequence. An example is provided by the

Fibonacci[1] *sequence*, in which each term $F(n)$ from the third onwards is the sum of the two immediately preceding terms:

$$F(1) = 1, F(2) = 1$$

$$F(n) = F(n-2) + F(n-1) \quad (n > 2)$$

It is left to you to check that the first few terms of the Fibonacci sequence are:

$$1, 1, 2, 3, 5, 8, 13, 21, \ldots$$

The Fibonacci sequence is a sequence for which there is a straightforward recursive definition, while the non-recursive definition is fairly complicated (see Exercise 8 at the end of this chapter).

Recursion can be used to define new functions. One important example is the factorial function, which we will need later when we study combinatorics.

Definition

The *factorial* of a natural number n is the product of all the natural numbers from 1 to n, and is written $n!$. Thus:

$$n! = n(n-1)(n-2) \ldots \times 3 \times 2 \times 1$$

We also define $0! = 1$.

For example, $4! = 4 \times 3 \times 2 \times 1 = 24$. The first few terms of the sequence of factorials (beginning with the 'zeroth' rather than the first term) are:

$$1, 1, 2, 6, 24, 120, 720, 5040, \ldots$$

Factorials arise in problems involving permutations and combinations. We will study such problems in Chapter 9; for the moment, the factorial function is just an example of a function having a simple recursive definition, which we are now about to find.

In order to find a recursive definition of $n!$, we begin by asking the following question: *how can $n!$ be calculated from* $(n-1)!$? For example, if we already know that $4! = 24$, how can we use that fact to evaluate $5!$? The answer is simple — since $4! = 4 \times 3 \times 2 \times 1$ and $5! = 5 \times 4 \times 3 \times 2 \times 1$, all we need to do is to multiply 5 by $4!$.

In general, we can evaluate $n!$ when $n > 0$ by multiplying n by $(n-1)!$. (Notice that this is true in particular when $n = 1$, but only because $0!$ is by definition equal to 1. This is one reason why $0!$ is defined in the way it is.) Having made these observations, we can now write down the recursive definition of $n!$:

---

[1]'Fibonacci' was the nickname of Leonardo of Pisa (1175–1250), who investigated this sequence in a problem involving the breeding of rabbits.

$$0! = 1$$

$$n! = n(n-1)! \quad (n > 0)$$

Notice that the recursion starts at $n = 0$ rather than $n = 1$.

We can check that this definition is correct for the first few values of n:

$$1! = 1 \times 0! = 1 \times 1 = 1$$

$$2! = 2 \times 1! = 2 \times 1 = 2$$

$$3! = 3 \times 2! = 3 \times 2 = 6$$

$$4! = 4 \times 3! = 4 \times 6 = 24$$

and so on.

Here is an iterative algorithm to evaluate $n!$, based on the recursive definition:

1. Input n
2. $f \leftarrow 1$
3. **For** $i = 1$ **to** n **do**
   3.1. $f \leftarrow i \times f$
4. Output f

Recursive definitions are not restricted to mathematical functions such as the factorial function. Here are recursive definitions of two concepts we have met in earlier chapters:

- Recursive definition of a bit string:
  1. 0 and 1 are bit strings.
  2. If s is a bit string then $s0$ and $s1$ are bit strings.

- Recursive definition of a logical expression:
  1. Any logical variable (denoted by a letter such as p) is a logical expression.
  2. If P and Q are logical expressions, then $\neg P$, $P \wedge Q$, $P \vee Q$, $P \rightarrow Q$ and $P \leftrightarrow Q$ are logical expressions.

These definitions are recursive because the term being defined ('bit string' in the first example and 'logical expression' in the second) appears within the definition itself.

EXAMPLE 7.1.4

| Show that 1001 is a bit string according to the recursive definition.

SOLUTION

1 is a bit string.   (Rule 1 of the definition)
Therefore 10 is a bit string.   (Rule 2 with $s = 1$)
Therefore 100 is a bit string.   (Rule 2 with $s = 10$)
Therefore 1001 is a bit string.   (Rule 2 with $s = 100$)

7.2 *Proof by induction*

Induction is basically a method for proving theorems about the natural numbers, although, as we will see, it can be applied far more broadly than this statement might at first suggest. In addition to being a powerful mathematical technique, induction is important in computing applications; it is closely related to recursion, and it is a useful tool if you are trying to establish that an algorithm is correct.

If we are going to attempt to prove a theorem using induction, the theorem must take the form of a statement that something is true for all the natural numbers. An example of such a statement is the following:

$$'1 + 2 + 3 + \ldots + n = \frac{n(n + 1)}{2} \text{ for all natural numbers } n'$$

We can check the claim made in this statement for any particular value of n by substituting that value into the formula and checking that the two sides are equal. For example, if $n = 6$ then the left-hand side is $1 + 2 + 3 + 4 + 5 + 6$ $= 21$ and the right-hand side is $\frac{6(6 + 1)}{2} = 21$.

Let's try another value, $n = 2$. The notation $1 + 2 + 3 + \ldots + n$ is a bit misleading here; the last term of the sum is 2 in this case, so the sum never gets to 3, despite the fact that '3' appears in the expression. The left-hand side is $1 + 2 = 3$ and the right-hand side is $\frac{2(2 + 1)}{2} = 3$.

The notational difficulty can be avoided by using 'summation notation'. In this notation, a sum of terms is indicated by means of the symbol Σ (upper-case Greek letter sigma). Using summation notation, we can write $1 + 2 + 3 + \ldots + n$ as $\sum_{i=1}^{n} i$, which we may read as 'the sum of i, as i goes from 1 to n'. We will not use summation notation in the remainder of this chapter, except in two of the exercises (Exercises 13 and 14).

Substituting values for n is rather like constructing a trace table for an algorithm — while doing a few checks like this might boost our confidence that the statement is correct, no amount of substituting particular values for n will prove that the formula is true for *all* values of n.

In order to prove the statement, we need to find a logical argument that establishes that the formula is always correct, regardless of what value n takes. Proof by induction provides a way of doing this.

A proof by induction on n consists of two steps:

1. In the *base step*, prove that the statement is true when $n = 1$.
2. In the *inductive step*, prove that if the statement is true for any particular value of n, then it is also true for the next value of n. In other words, prove that if the statement is true when $n = k$, then it is also true when $n = k + 1$, where k stands for a fixed but arbitrary value that n can take.

If these two steps can be carried out, then we can conclude that the statement must be true for all natural number values of n, by reasoning in the following way:

The statement is certainly true when $n = 1$, because the base step shows this.

Now, because the statement is true when $n = 1$, we may put $k = 1$ in the inductive step, from which it follows that the statement must also be true when $n = 2$.

Now that we know that the statement is true when $n = 2$, we can deduce that it must also be true when $n = 3$, by applying the inductive step with $k = 2$.

Stepping up through the natural numbers in this way, and repeatedly applying the inductive step, we conclude that the statement must be true successively for $n = 4$, $n = 5$, $n = 6$, and in general for all natural number values of n.

The induction process is rather like a row of dominoes falling. Imagine a large number of dominoes (in theory, an infinite number) standing on their ends in a row on a table. You knock the first domino, causing it to fall; this corresponds to the base step. As each domino falls, it causes the next one to fall; this corresponds to the inductive step. The first domino knocks over the second one, which knocks over the third, which knocks over the fourth, and so on. Eventually, every domino will fall.

EXAMPLE 7.2.1

Prove by induction: $1 + 2 + 3 + \ldots + n = \dfrac{n(n + 1)}{2}$ for all natural number values of n.

SOLUTION

Base step: When $n = 1$, the left-hand side equals 1 and the right-hand side equals $\dfrac{1(1 + 1)}{2} = 1$, so the formula is true in this case.

Inductive step: We begin the inductive step by assuming that the formula is true when n equals some fixed but arbitrary natural number k:

$$1 + 2 + 3 + \ldots + k = \frac{k(k + 1)}{2}$$

(This assumption is called the *inductive hypothesis*.)

We need to *deduce* from this assumption that the formula is also true when $n = k + 1$. Now, if $n = k + 1$, then the left-hand side of the formula is:

$$1 + 2 + 3 + \ldots + k + (k + 1)$$

where we have explicitly written the second-last term (k) as well as the last one ($k + 1$), in order to make the next step of the proof clearer.

We can now use the inductive hypothesis to replace $1 + 2 + 3 + \ldots + k$ in this expression by $\dfrac{k(k + 1)}{2}$:

$$1 + 2 + 3 + \ldots + k + (k + 1) = \frac{k(k + 1)}{2} + (k + 1)$$

In order to complete the inductive step, it remains for us to show that $\dfrac{k(k + 1)}{2} + (k + 1)$ is equal to the right-hand side of the formula with n replaced by $k + 1$. This can be done in a few lines of algebra:

$$\frac{k(k + 1)}{2} + (k + 1) = \frac{(k + 1)k + (k + 1)2}{2}$$

$$= \frac{(k + 1)(k + 2)}{2}$$

$$= \frac{(k + 1)[(k + 1) + 1]}{2}$$

where the last line is just $\dfrac{n(n + 1)}{2}$ with $k + 1$ in place of n.

We began the inductive step by assuming that the formula is true when $n = k$. We then deduced from that assumption that the formula is true when $n = k + 1$. In short, we have just shown that if the formula is true when $n = k$, then it is true when $n = k + 1$. This completes the inductive step.

To conclude the proof, it remains only to say something like this: by induction, $1 + 2 + 3 + \ldots + n = \dfrac{n(n + 1)}{2}$ for all natural numbers n.

Notice that we did not actually *derive* the formula $1 + 2 + 3 + \ldots + n = \dfrac{n(n + 1)}{2}$. For all we can tell, it might have been obtained using a logical argument, but it could just as well have been found by making a lucky guess. Once we have written down a result that we suspect to be true, however, we can then use induction to prove that it really is true. This is simply the nature of induction; by itself it cannot be used to derive new theorems, but it can be used to prove that a conjecture obtained by other means is correct.

The solution to Example 7.2.1 might appear rather long-winded, but that is only because every step has been explained in detail. The next example shows how a proof by induction is usually set out in practice. (It also illustrates the fact that a statement to be proved by induction need not be a formula.)

EXAMPLE 7.2.2

Prove by induction: $n^2 + n$ is an even number, for every natural number n.

SOLUTION

Base step: If $n = 1$ then $n^2 + n = 1^2 + 1 = 2$, which is even.
Inductive step: Assume $k^2 + k$ is even. Then:

$$(k + 1)^2 + (k + 1) = k^2 + 2k + 1 + k + 1$$
$$= k^2 + k + 2k + 2$$

$k^2 + k$ is even, by the inductive hypothesis, and $2k + 2$ is clearly also even. Therefore their sum, $k^2 + k + 2k + 2$, must be even.

By induction, $n^2 + n$ is an even number, for every natural number n.

When once you have completed the base step and the inductive step, the 'domino effect' ensures that the result is true when $n = 1, 2, 3, 4$, and so on. In the proof, there is no need to say: 'It's true for $n = 1$, therefore it's true for $n = 2$, therefore it's true for $n = 3$, and so on.' All of that is summed up in the phrase: 'By induction'.

Notice in the last example that the expression in the inductive step was simplified to $k^2 + k + 2k + 2$ rather than $k^2 + 3k + 2$, so that the inductive hypothesis could be applied. When you are carrying out the inductive step, it is helpful to keep in mind that the inductive hypothesis must be used at some stage, as this will have some bearing on the steps you need to perform.

Constructing proofs by induction takes practice. While the details of the proof will vary from one problem to another, there are a number of useful guidelines that should be kept in mind, because they apply to any proof by induction. We list them below:

1. Every proof by induction must have a base step and an inductive step.
2. The base step is usually quite easy to construct. Sometimes it might even seem trivial, but it is still an essential part of the proof, and must always be included.
3. Begin the inductive step with the inductive hypothesis. Use the word 'Assume', and replace the variable (n in our examples) with a fixed but arbitrary value (k).
4. Complete the inductive step by proving that the result is true when n is replaced by $k + 1$. The inductive hypothesis *must* be used at some stage in the reasoning.
5. Finish the proof with a statement along the lines of: 'By induction, ... for every natural number n'.

7.3 *Induction and recursion*

You have probably noticed by now that essentially the same idea underlies both recursion and induction. A recursive definition consists of a base case (usually $n = 1$), and a recursive formula for successively obtaining each case ($n = 2, 3, 4, \ldots$) from the previous one. In a similar way, we carry out a proof by induction by proving the base case ($n = 1$), and then showing in the inductive step that each case from 2 onwards can be deduced from the previous one. We will now make this connection explicit by showing how induction can be used to prove a non-recursive formula for the nth term of a sequence if a recursive definition is given.

EXAMPLE 7.3.1

A sequence is defined recursively as follows:

$$t(1) = 1$$

$$t(n) = t(n - 1) + 2n - 1 \quad (n > 1)$$

Conjecture a non-recursive formula for $t(n)$, and use induction to prove that the conjecture is correct.

SOLUTION

In order to make a conjecture about a formula for $t(n)$, we first need to evaluate some of the terms of the sequence:

$$t(1) = 1$$

$$t(2) = t(1) + 2 \times 2 - 1 = 1 + 2 \times 2 - 1 = 4$$

$$t(3) = t(2) + 2 \times 3 - 1 = 4 + 2 \times 3 - 1 = 9$$

Continuing in this way, we obtain the sequence:

$$1, 4, 9, 16, 25, 36, 49, \ldots$$

The pattern of the numbers suggests (but does not prove) that $t(n) = n^2$. This is our conjecture, which we now prove by induction.

Base step: $t(1) = 1 = 1^2$, so the conjecture is true when $n = 1$.

Inductive step: Assume that $t(k) = k^2$ for some (fixed but arbitrary) natural number k. Then:

$$\begin{aligned}
t(k + 1) &= t(k) + 2(k + 1) - 1 \quad \text{(by the definition of } t(n) \text{ with } n = k + 1) \\
&= k^2 + 2(k + 1) - 1 \quad \text{(by the inductive hypothesis)} \\
&= k^2 + 2k + 1 \\
&= (k + 1)^2
\end{aligned}$$

that is, $t(n) = n^2$ when $n = k + 1$.

By induction, $t(n) = n^2$ for every natural number n.

We saw earlier how the recursive and non-recursive definitions of a sequence can form the basis of different algorithms for generating the terms of the sequence. It can be useful to be able to prove that the two definitions are equivalent if we want to establish that an algorithm of this type is producing the correct output.

For example, the following algorithm accepts a natural number n as input and produces a table of cubes from 1 to n as output:

    1. Input n
    2. *icubed* $\leftarrow$ 0
    3. **For** $i = 1$ **to** n **do**
        3.1. $j \leftarrow 3i(i-1) + 1$
        3.2. *icubed* $\leftarrow$ *icubed* $+ j$
        3.3. Output i, *icubed*

How can we be sure that the algorithm is correct? We could carry out a trace for some particular inputs, but what is really needed is a general argument to show that the algorithm works correctly with any (valid) input.

What we need to show is that *icubed* equals the cube of i each time Step 3.3 is executed. Suppose we denote the value of *icubed* when Step 3.3 is being executed for the ith time by *icubed*(i). We want to show that *icubed*(i) = i^3 for all values of i from 1 to n. We can do this in two steps:

1. Write down the recursive formula for *icubed*(i) on which the algorithm is based.
2. Prove that *icubed*(i) = i^3 using induction on i.

There is no problem with using induction here, even though we only want to prove the result for $i \leq n$, rather than for all natural number values of i. The inductive step will be valid only for values of k ranging from 1 to $n - 1$, but that is sufficient to ensure that the 'domino effect', on which induction depends, will apply to all the values of i we are interested in, up to and including $i = n$. It is as if we had a row of just n dominoes to knock over, rather than an infinite number.

Look at how *icubed*(i) is calculated in the algorithm. In Step 3.2, it is obtained by adding j to the previous value of *icubed*, namely *icubed*($i - 1$). The value of j was calculated in Step 3.1 as $3i(i-1) + 1$. Putting these two steps together, we obtain the following recursive formula for *icubed*:

$$icubed(i) = icubed(i - 1) + 3i(i - 1) + 1$$

This formula is valid when i lies in the range from 2 to n, but not if $i = 1$, because *icubed*(0) has no meaning. In order to find *icubed*(1), we have to step through the algorithm from the beginning to the point where Step 3.3 is being executed for the first time. It is easy to check that *icubed*(1) = 1, and this forms the base for the recursion.

It remains now to carry out the following task. Let *icubed* be defined recursively as follows:

$$icubed(1) = 1$$

$$icubed(i) = icubed(i - 1) + 3i(i - 1) + 1 \quad (2 \leq i \leq n)$$

Prove that $icubed(i) = i^3$ for all values of i such that $1 \leq i \leq n$.

Here is the proof, using induction on i:

Base step: $icubed(1) = 1 = 1^3$, so the result is true when $i = 1$.
Inductive step: Assume $icubed(k) = k^3$ for a fixed value of k such that $1 \leq k \leq n - 1$. Then:

$$
\begin{aligned}
icubed(k + 1) &= icubed(k) + 3(k + 1)k + 1 \\
&= k^3 + 3(k + 1)k + 1 \\
&= k^3 + 3k^2 + 3k + 1 \\
&= (k + 1)^3
\end{aligned}
$$

By induction, $icubed(i) = i^3$ for all values of i such that $1 \leq i \leq n$.

A word of caution is appropriate at this point. The practical problem of finding a method for proving the correctness of a given computer program is enormous (and in fact is known not to be solvable, in general). We have only looked at one very small aspect of the problem here. Quite apart from the fact that the method we have presented here can be applied only to a very restricted class of algorithms, we have not considered practical questions such as whether the algorithm has been implemented correctly in a programming language, and whether any problems might arise from the particular ways in which computers store and manipulate data, to name just two areas where difficulties could occur.

7.4 *Recursively defined functions and recursive algorithms*

Earlier in this chapter, we introduced the factorial function and saw how it could be defined recursively. As we saw in Chapter 6, many programming languages provide the facility for defining new functions, so it is natural to ask whether recursively defined functions are permissible in programming languages. The short answer is that recursive definitions are allowed in some languages (Pascal, C and Lisp are examples), but not in others. Before we investigate recursion in programming languages, however, we need to look first at how functions are handled in pseudocode.

A user-defined function in a programming language is often defined in a function subprogram. The subprogram forms a self-contained module within a program, and is 'called' or referenced from another part of the program. A function subprogram is similar in some ways to an entire program, but there are some important differences. Whereas a program has input from the 'real world' (via a keyboard or other input device), the input to a function subprogram comes from the program module that calls it, and is passed to the subprogram via a parameter (or a list of parameters). Similarly, the output from a function subprogram does not go to an output device such as a screen, but is returned to the calling module via the name of the function.

These differences affect the way in which we write algorithms for functions in pseudocode. Instead of using input and output statements, we will begin the code with a function header containing the name of the function and an input parameter (rather like the Pascal function header that was given as an example in Section 6.3). With this arrangement, we will not need an input statement at all, and an assignment statement (using the name of the function) will take the place of an output statement.

As an illustration of these ideas, suppose we want to write an algorithm to evaluate $n!$ as a function subprogram. Earlier in this chapter we wrote an algorithm for evaluating $n!$, so we can use that here with just a few minor changes:

Function *factorial*(n):
1. $f \leftarrow 1$
2. **For** $i = 1$ **to** n **do**
    2.1. $f \leftarrow i \times f$
3. *factorial* $\leftarrow f$

We can imagine that there is another algorithm that 'calls' this one with a particular value of n, perhaps using a statement like this:

$$x \leftarrow factorial(5)$$

The value $n = 5$ serves as the input to the function. The value assigned to *factorial* in Step 3 (5! = 120 in this example) would replace *factorial*(5) in the calling algorithm, so that the value 120 would be assigned to x. Note that Step 3 differs from an ordinary assignment statement, in that the identifier *factorial* to the left of the arrow represents a function, not a variable. In a sense, Step 3 is really more like an output statement, although we have written it in the way we have in order to be consistent with the syntax of programming languages such as Pascal.

The algorithm we have constructed here is an iterative rather than a recursive algorithm, although it is clearly based on the recursive definition of $n!$. Suppose, however, that we now rewrite the algorithm in a way that resembles the recursive definition even more closely. After all, the recursive definition doesn't immediately suggest the use of a **For-do** loop; it really looks rather more like an **If-then-else** — if $n = 0$ then $n! = 1$, otherwise $n! = n(n-1)!$. We can translate this directly into pseudocode in an obvious way:

Function *factorial*(n):
1. **If** $n = 0$ **then**
    1.1. *factorial* $\leftarrow 1$
    else
    1.2. *factorial* $\leftarrow n \times$ *factorial*($n - 1$)

This is no longer just an algorithm based on a recursive formula; it is a *recursive algorithm*, because it contains a call to *itself*.

Let's see how an algorithm written in this way would work. For example, suppose we want to use the algorithm to evaluate 2!. Here is what happens:

First, Step 1 is entered with $n = 2$. The condition $n = 0$ is not satisfied, so we move to Step 1.2.

We are now faced with the task of evaluating $n \times factorial(n - 1)$ with $n = 2$, that is, $2 factorial(1)$. Before we can do this, we need to put the execution of the algorithm 'on hold' while we use the same algorithm to evaluate $factorial(1)$. It is helpful to imagine that a copy of the algorithm is to be used to evaluate $factorial(1)$, since the original algorithm is 'busy'.

With execution of the first copy of the algorithm suspended, we enter the second copy of the algorithm with $n = 1$. Again we go to Step 1.2, and again we must place the algorithm on hold, this time while $factorial(0)$ is being evaluated.

Now, with two copies of the algorithm on hold, we enter a third copy with $n = 0$. This time, we go to Step 1.1 and follow the algorithm through to completion, returning a value of 1.

We can now return to the second copy, which we had suspended while we evaluated $factorial(0)$. We substitute the value we have obtained for $factorial(0)$ into Step 1.2 to get 1×1, and then evaluate this expression to obtain 1 for the value of $factorial(1)$.

Finally, returning to the first copy of the algorithm with the value we have found for $factorial(1)$, we can finish the computation to obtain $2 \times 1 = 2$ as the value of $factorial(2)$.

EXAMPLE 7.4.1

Write a recursive algorithm for the sequence in Example 7.1.2 regarded as a function.

SOLUTION

Here is the recursive definition of the sequence in Example 7.1.2:

$$t(1) = 3$$

$$t(n) = 2t(n - 1) + n \quad (n > 1)$$

Recall that a sequence is really just a particular type of function. We can therefore write a recursive algorithm for this sequence as a function:

Function $t(n)$:
1. **If** $n = 1$ **then**
    1.1. $t \leftarrow 3$
  else
    1.2. $t \leftarrow 2t(n - 1) + n$

Let us suppose now that we want to write an algorithm to output the first m terms of the sequence in the example above. This is the same problem as in Example 7.1.3, except that we would now like to solve it using recursion.

One way of doing this is to write an algorithm that calls the function we have just written:

   1.   Input m
   2.   **For** $n = 1$ **to** m **do**
       2.1. $x \leftarrow t(n)$
       2.2.   Output x

Although this works, it is not very efficient computationally. Each time the **For-do** loop is executed, $t(n)$ is calculated recursively from $t(n-1)$; but in order to do this, $t(n-1)$ must first be calculated recursively, despite the fact that it has already been evaluated during the previous pass through the loop.

There is another way — we can write the entire algorithm recursively. Up to this point, the only recursive algorithms we have met are function subprograms, but the technique of recursion is actually much more general. In fact, recursion represents a whole new way of approaching the problem of designing an algorithm.

The key to using the recursive approach to a problem is to work out how to solve the problem *assuming that a simpler case of the same problem has already been solved*. By reducing the problem to a simpler one in this way, we initiate a chain in which each case is reduced in turn to a simpler one, until we reach a base case for which the problem can be solved very easily.

In order to apply recursion to the present problem, we need to answer the following question: How could we output the first m terms of the sequence, *if we already had an algorithm to output the first $m-1$ terms?*

If $m > 1$, the answer is:

   1.   Output the first $m-1$ terms (using the algorithm we already have).
   2.   Evaluate $t(m)$.
   3.   Output the value of $t(m)$.

(For the base case $m = 1$, we simply omit Step 1.)

There is one question that needs to be answered before we can write this algorithm in pseudocode: how is Step 2 to be carried out? We could call the recursive algorithm we have written for $t(m)$, but that would lead to precisely the inefficiency we are trying to avoid. What we would like to do in Step 2 is to evaluate $t(m)$ using the value of $t(m-1)$ that has already been obtained in Step 1. In order to be able to do this, the algorithm to output the first $m-1$ terms will need to return the value of $t(m-1)$ to the algorithm that calls it.

Here is the algorithm written in pseudocode. We need to name the algorithm so that it can be called, and the number of terms to be output (m) and the last (i.e. the mth) term (*last_term*) must be listed as parameters. (Think of m as the 'input' and *last_term* as the 'output' of the algorithm.)

Algorithm *write_sequence(m, last_term)*:
1. **If** $m = 1$ **then**
   1.1. *last_term* $\leftarrow 3$
   else
   1.2. *write_sequence(m – 1, secondlast_term)*
   1.3. *last_term* $\leftarrow 2$ *secondlast_term + m*
2. Output *last_term*

The name *secondlast_term* refers to the fact that it is the second-last term of the sequence with m terms (and therefore the last term of the sequence with $m - 1$ terms). In Step 1.2, the algorithm is called with the *formal* parameters m and *last_term* replaced by the *actual* parameters $m - 1$ and *secondlast_term*.

Algorithm *write_sequence* cannot stand alone; it must be called from another algorithm. This can be done very simply:

1. Input m
2. *write_sequence(m)*

EXAMPLE 7.4.2

| Trace Algorithm *write_sequence* when $m = 3$.

SOLUTION

Because the algorithm is recursive, we will need to keep track of the values of the variables separately in each copy of the code.

Step	Copy 1			Copy 2			Copy 3			Output
	m	*last*	*second-last*	m	*last*	*second-last*	m	*last*	*second-last*	
Initially 1.2 (hold)	3	–	–							
Initially 1.2 (hold)				2	–	–				
Initially							1	–	–	
1.1							1	3	–	
2							1	3	–	3
1.2				2	–	3				
1.3				2	8	3				
2				2	8	3				8
1.2	3	–	8							
1.3	3	19	8							
2	3	19	8							19

Table 7.1

The process of constructing a recursive algorithm may be new to you, so we give another example.

EXAMPLE 7.4.3

Construct a recursive algorithm to find the smallest number in a list of numbers.

SOLUTION

Let $x_1, x_2, \ldots, x_n$ be the list of numbers. Suppose we already had an algorithm to input the list $x_1, x_2, \ldots, x_{n-1}$ consisting of the first $n-1$ of the numbers, and find the smallest number in the list. We could then proceed in the following way (when $n > 1$):

1. Find the smallest number, *min*, in the first $n-1$ numbers.
2. Input x_n
3. **If** $x_n < min$ **then**
   3.1. $min \leftarrow x_n$

The final value of *min* is the smallest number in the list.

Let *find_minimum* be the name of the algorithm. The input parameter is n, and the output parameter is the minimum of the first n numbers. In order to handle the case $n = 1$ correctly, it is convenient to input x_n first. The recursive algorithm can be written as follows:

Algorithm *find_minimum*(n, *min*):
1. Input x_n
2. **If** $n = 1$ **then**
   2.1. $min \leftarrow x_1$
   else
   2.2. *find_minimum*($n - 1$, *min*)
   2.3. **If** $x_n < min$ **then**
       2.3.1. $min \leftarrow x_n$

7.5 *Recursively defined functions in programming languages*

In order to relate this material to actual programming languages, we will conclude this chapter by looking briefly at recursive algorithms for functions in Pascal and the Lisp-based language PC-Scheme®. We will use as our example the sequence defined in Example 7.3.1:

$$t(1) = 1$$

$$t(n) = t(n-1) + 2n - 1 \quad (n > 1)$$

The terms of this sequence can be generated using a recursive algorithm for $t(n)$, which looks like this in pseudocode:

Function $t(n)$:
1. **If** $n = 1$ **then**
   1.1. $t \leftarrow 1$
  else
   1.2. $t \leftarrow t(n-1) + 2n - 1$

In Pascal, both the type of the input parameter and the type of the value of the function must be declared in the function header. In order to keep the code consistent with the mathematical definition of a sequence as a function from **N** to **R**, we will declare the type of the function to be real, even though all the terms of this particular sequence are integers.

Here is the Pascal code:

```
function t(n: integer): real;
begin
   if n = 1 then
      t := 1.0
   else
      t := t(n-1) + 2.0 * n - 1.0
end;
```

It is not surprising that the Pascal code looks very similar to the pseudocode version of the algorithm, since the pseudocode we are using is based closely on Pascal syntax.

The PC-Scheme version looks a bit different:[2]

```
(define t
   (lambda (n)
   (if (= n 1)
   1
   (- (+ (t (- n 1)) (* 2 n)) 1)))))
```

The code takes the form of a list of three items written between parentheses: the operator `define`, the name `t` of the function, and the definition of the function. The definition of the function is itself a list of two items: the first, `lambda (n)`,[3] indicates that a rule for a function expressed in terms of a variable n is about to follow, and the second is a list that gives that rule. In turn, some of the items in the rule are lists, and so on. PC-Scheme uses prefix

---

[2]If you have programmed in Lisp, you might object that t has a special meaning, and should not be used as the name of a function. This is true in some versions of Lisp, but not in PC-Scheme.

[3]'Lambda' is the name of the Greek letter written λ, which is used as a symbol in the branch of logic known as lambda calculus, with a similar meaning to its meaning in PC-Scheme.

notation for operations (such as addition and subtraction) and relations (such as equality). For example, $n - 1$ is written $(-\ n\ 1)$, and $n = 1$ is written $(=\ n\ 1)$. With this explanation, you should be able to follow the gist of the code, even if you are not familiar with PC-Scheme.

Recursion can be a powerful way of approaching problems in programming, especially in languages like Lisp, which are designed to take full advantage of it. This chapter has provided an introduction to induction and recursion; we will make use of both of these tools again in the chapters to follow.

> ## Exercises

1. Evaluate $t(2)$, $t(3)$ and $t(4)$ for the following recursively defined sequences:

   (a) $t(1) = 3$
   $t(n) = t(n - 1) + 4$ $(n > 1)$
   (b) $t(1) = 1$
   $t(n) = t(n - 1) + n$ $(n > 1)$
   (c) $t(1) = 0$
   $t(n) = 2t(n - 1) + 1$ $(n > 1)$
   (d) $t(1) = 2$
   $t(n) = t(n - 1) + n^2$ $(n > 1)$

2. Evaluate $t(3)$, $t(4)$ and $t(5)$ for the sequence:

$$t(1) = 1,\ t(2) = 1$$

$$t(n) = t(n - 1) + 2t(n - 2)   (n > 2)$$

3. For each of the sequences in Exercises 1(a) and 1(b), write an iterative algorithm to output the first m terms of the sequence.

4. Write down a recursive definition for 2^n (where n is a non-negative integer), that is, give the value of 2^0, and give a formula for 2^n in terms of 2^{n-1} when $n > 0$.

5. Let $t(n) = 1 + 3 + 5 + \ldots + (2n - 1)$. Write down a recursive definition of $t(n)$.

6. Find a recursive and a non-recursive definition for the sequence:

$$2, 7, 12, 17, 22, 27, \ldots$$

7. The first term of a certain sequence is 1, and each subsequent term is obtained by doubling the previous term and adding 3 to the result. Write down a recursive definition of the sequence.

8. The non-recursive formula for $F(n)$, the nth term of the Fibonacci sequence, is shown below:

$$F(n) = \frac{1}{\sqrt{5}} \left[\left(\frac{1 + \sqrt{5}}{2} \right)^n - \left(\frac{1 - \sqrt{5}}{2} \right)^n \right]$$

Verify that the non-recursive formula given for the Fibonacci sequence is correct for the first few values of n.

9. Without using a calculator, and using the properties of factorials to facilitate the calculations as far as possible, simplify or evaluate the following expressions:

(a) $\dfrac{25!}{24!}$

(b) $\dfrac{11!}{8!}$

(c) $\dfrac{10!}{7!\ 3!}$

(d) $\dfrac{n!}{(n-2)!}$

10. Construct a trace table for the iterative factorial algorithm given in the text, for the case $n = 4$.

11. Use the recursive definition given in the text to show that $\neg(p \wedge q) \vee q$ is a logical expression.

12. Prove by induction that the following statements are true for all natural numbers n:

(a) $1 + 3 + 5 + \ldots + (2n - 1) = n^2$

(b) $1^2 + 2^2 + 3^2 + \ldots + n^2 = \dfrac{n(n + 1)(2n + 1)}{6}$

(c) $n^3 - n$ is divisible by 3

(d) $5^n - 1$ is divisible by 4

(e) $\frac{1}{3} + \frac{1}{9} + \frac{1}{27} + \ldots + \frac{1}{3^n} = \dfrac{1 - \frac{1}{3^n}}{2}$

13. The summation notation (mentioned briefly earlier in this chapter) is defined as follows. Suppose that $f(i)$ is a general formula for the ith term in the sum. Then:

$$\sum_{i = m}^{n} f(i) = f(m) + f(m + 1) + f(m + 2) + \ldots + f(n)$$

Examples:

- $\displaystyle\sum_{i = 1}^{5} i^2 = 1^2 + 2^2 + 3^2 + 4^2 + 5^2 = 55$

- $\displaystyle\sum_{j = 4}^{7} (2j + 1) = (2 \times 4 + 1) + (2 \times 5 + 1) + (2 \times 6 + 1) + (2 \times 7 + 1)$

  $= 48$

Evaluate the following sums:

(a) $\displaystyle\sum_{i = 1}^{4} (3i - 2)$

(b) $\displaystyle\sum_{k = 1}^{6} k(k + 4)$

(c) $\displaystyle\sum_{i = 5}^{9} (i^2 - i)$

14. Prove by induction:

(a) $\sum_{i=1}^{n} (4i - 1) = n(2n + 1)$

(b) $\sum_{i=1}^{n} 2^{i-1} = 2^n - 1$

15. Let $t(n)$ be defined recursively as follows:

$$t(1) = 2$$
$$t(n) = t(n - 1) + 3n - 1 \quad (n > 1)$$

Prove by induction that $t(n) = \dfrac{n(3n + 1)}{2}$

16. Let $t(n)$ be defined recursively as follows:

$$t(1) = 1$$
$$t(n) = t(n - 1) + n \times n! \quad (n > 1)$$

Prove by induction that $t(n) = (n + 1)! - 1$

17. For each of the following recursively defined sequences, evaluate the first few terms, conjecture a formula for $t(n)$, and use induction to prove that the formula is correct:

(a) $t(1) = 3$

$$t(n) = t(n - 1) + 5 \quad (n > 1)$$

(b) $t(1) = 1$

$$t(n) = t(n - 1)\left(1 - \frac{1}{n}\right) \quad (n > 1)$$

(c) $t(1) = 2$

$$t(n) = t(n - 1) + 2n - 1 \quad (n > 1)$$

18. Suppose we are able to prove the following results for a statement about any natural number n:

1. The statement is true when $n = 5$.
2. For any natural number k, if the statement is true when $n = k$ then it is true when $n = k + 1$.

For which natural numbers could we be certain that the statement is correct?

19. Suppose we are able to prove the following results for a statement about any natural number n:

1. The statement is true when $n = 1$.
2. For any natural number k, if the statement is true when $n = k$ then it is true when $n = k + 2$.

For which natural numbers could we be certain that the statement is correct?

20. Using a minor variation of proof by induction in the form in which we have presented it, it is possible to prove results about all *non-negative* integers n. This is done by using $n = 0$ instead $n = 1$ in the base step.

    Prove the following results by induction:

    (a) $\dfrac{6^n - 1}{5}$ is an integer for $n \geq 0$

    (b) $2^n > n$ for $n \geq 0$

21. Find a recursive expression for the output of each of the following algorithms. Hence prove by induction that the algorithm produces the output claimed.

    (a) Algorithm to output a table of values of $\dfrac{i(i + 3)}{2}$, $i = 1, 2, 3, \ldots, n$:

    1. Input n
    2. *value* $\leftarrow 0$
    3. **For** $i = 1$ **to** n **do**
       3.1. $j \leftarrow i + 1$
       3.2. *value* $\leftarrow$ *value* $+ j$
       3.3. Output i, *value*

    (b) Algorithm to output a table of values of $\dfrac{i(2i + 1)(2i - 1)}{3}$, $i = 1, 2, 3, \ldots, n$:

    1. Input n
    2. *value* $\leftarrow 0$
    3. **For** $i = 1$ **to** n **do**
       3.1. $j \leftarrow (2i - 1)^2$
       3.2. *value* $\leftarrow$ *value* $+ j$
       3.3. Output i, *value*

22. Write a recursive algorithm for each of the sequences in Exercises 1(a) and 1(b) regarded as functions.

23. For each of the sequences in Exercises 1(c) and 1(d), write a recursive algorithm to output the first m terms of the sequence.

24. Trace the algorithms in Exercise 23 with $m = 3$.

25. Write a recursive algorithm to test whether a sequence of n bits consists entirely of zeros.

26. Write a recursive algorithm to count the number of times the letter 'a' occurs in a string of n characters $c_1 c_2 \ldots c_n$.

27. The algorithm shown below generates the terms of the Fibonacci sequence:

Function *fibonacci(n)*:
1. **If** $n = 1$ or $n = 2$ **then**
    1.1.   *fibonacci* $\leftarrow 1$
    else
    1.2.   *fibonacci* $\leftarrow$ *fibonacci*$(n - 2)$ + *fibonacci*$(n - 1)$

Although this algorithm is correct, it is very inefficient. Explain why this is so. (Hint: Carrying out a trace for a particular value of n may help to locate the problem.)

8. BOOLEAN ALGEBRA AND DIGITAL CIRCUITS

8.1 *Boolean algebra*

In this chapter we will take a look at the branch of mathematics known as Boolean algebra. There are two main reasons for studying Boolean algebra at this point. Firstly, we will be able to see how Boolean algebra draws together into a unified theory many of the concepts in propositional logic and sets that we met in Chapters 4 and 5. The idea of incorporating two (or more) separate topics into a single theory is a powerful concept, which has played an important role in the development of mathematics. By identifying the common rules that underlie logic and sets, we will be able to derive results that can be applied to both areas.

The second reason for studying Boolean algebra is that the notation of Boolean algebra is widely used in computing, especially in the design of digital circuits. In the latter part of this chapter, we will see how Boolean algebra is used in designing the circuitry needed to perform digital computations.

Let us start by recalling the work we did in Chapter 4 on propositional logic. The objects we were working with there were *logical expressions*. When we used a variable such as p to denote a logical expression, it was natural to think of p as standing for a proposition, in the same way as we think of a variable x in algebra as standing for a number. When we do symbolic logic, however, we are not really working with the propositions themselves but with their truth values. Seen from this point of view, a logical expression such as p can only have two values: true (T) and false (F). So we see that it is really the 2-element set {T, F} that takes the place in logic that the set **R** of real numbers takes in ordinary algebra.

We can express this idea in another way using the concept of equivalence relations introduced in Chapter 5. The relation 'has the same truth value as' is reflexive, symmetric and transitive, so it is an equivalence relation on the set of all the propositions we are dealing with. There are two equivalence classes: the set of all the true propositions and the set of all the false propositions. We can think of T as standing for the first of these equivalence classes and F for the second.

In the same way as addition and multiplication are operations defined on the set **R** of real numbers, the logical connectives introduced in Chapter 4 are operations on the set {T, F}. In this chapter we will be concerned with just three of the connectives: **and**, **or** and **not**. The way each of these connectives operates is set out in its truth table, just as the way in which numbers are added could be defined by writing down an addition table. We saw in Chapter 4 that the logical operations obey certain laws, some of which are similar in form to laws of ordinary algebra.

In summary, propositional logic consists of:

- A set $\{T, F\}$.
- Three operations, $\wedge$, $\vee$ and $\neg$, defined on the set.
- A list of laws that the three operations satisfy.

Now suppose we put propositional logic aside for the moment and consider the power set $\mathcal{P}(A)$ of a set A. (Recall that $\mathcal{P}(A)$ means the set of all of the subsets of A.) The operations of intersection, union and complement are operations on $\mathcal{P}(A)$, because the intersection and union of any two elements of $\mathcal{P}(A)$ are also elements of $\mathcal{P}(A)$, and the complement of any element of $\mathcal{P}(A)$ is an element of $\mathcal{P}(A)$. In Chapter 5 we listed some laws that these three operations satisfy.

Summarising, here is what we have:

- A set $\mathcal{P}(A)$.
- Three operations, $\cap$, $\cup$ and $^-$, defined on the set.
- A list of laws that the three operations satisfy.

This looks very similar to what we had in propositional logic. In fact, we can say more — we saw in Chapter 5 that it is really the *same* list of laws in both cases, provided that we replace the logical connectives by their corresponding set operations.

This does not mean that logic and power sets are the same in every respect. The set we deal with in logic has just two elements: T and F. A power set $\mathcal{P}(A)$ will usually have more than two elements, although two of those elements, $\varnothing$ and A, play a special role, because they have a place in the laws of sets that corresponds to the place occupied by F and T in the laws of logic. What we *can* say is that logic and power sets have much in common; in particular, their operations obey essentially the same laws.

We can exploit the similarity between logic and power sets in the following way. Instead of studying logic or power sets, as we have been doing until now, we can study *the laws themselves*. By doing that, we will in a sense be studying logic and power sets simultaneously (and also any other mathematical system that satisfies those laws), because anything that can be deduced from the laws must be true when interpreted for logic, for power sets, or for any other system in which the laws apply.

We have been using the word 'operation' without having defined it properly. Here are the definitions of the two types of operations we will be needing.

Definitions

A *binary operation* on a set X is a function with domain $X \times X$ and codomain X.

A *unary operation* on X is a function with domain and codomain X.

A binary operation can be thought of as a rule that takes *two* elements of a set, and produces an element of the set as the answer. Addition is an example of a binary operation on **R**, because we can think of addition as a function that takes each element (x, y) of **R** $\times$ **R** and maps it to the element $x + y$ of **R**. In symbols, addition is a function f defined as follows:

$$f: \mathbf{R} \times \mathbf{R} \to \mathbf{R}, \quad f(x, y) = x + y$$

(Strictly, we should write $f((x, y))$ for the image of (x, y), but the notation $f(x, y)$ is usual.)

Subtraction and multiplication (but not division) are also binary operations on **R**. Similarly, the logical connectives **and** and **or** are binary operations on $\{T, F\}$; for example, we could express **and** as a function by writing:

$$f: \{T, F\} \times \{T, F\} \to \{T, F\}, \quad f(p, q) = p \wedge q$$

In the same way, intersection and union are binary operations on the power set $\mathcal{P}(A)$ of a set A.

In a similar manner, a unary operation is a rule that takes *one* element of a set and produces an element of the set as the answer. The process of negating a number is an example of a unary operation on **R**:

$$f: \mathbf{R} \to \mathbf{R}, \quad f(x) = -x$$

Somewhat confusingly, the minus sign is used with two different meanings in algebra. When we do a subtraction, $x - y$, we are using the minus sign to denote a *binary* operation, whereas when we negate a number, $-x$, we are using it to denote a *unary* operation. (On most calculators, the distinction between binary and unary minus is made explicit because they correspond to different keys.)

The connective **not** is a unary operation on $\{T, F\}$, and complementation is a unary operation on a power set.

Now we are ready for the main definition in this chapter.

Definition

A *Boolean algebra*[1] consists of a set B, two binary operations on B (denoted by + and ×), a unary operation (denoted by '), and two distinguished elements of B (denoted by 0 and 1), such that the following axioms are satisfied for all elements x, y and z of B:

- $x + y = y + x$ $\qquad$ $x \times y = y \times x$ $\qquad$ commutative axioms
- $x + (y + z) =$ $\qquad$ $x \times (y \times z) =$
  $(x + y) + z$ $\qquad\quad$ $(x \times y) \times z$ $\qquad$ associative axioms
- $x + (y \times z) =$ $\qquad$ $x \times (y + z) =$
  $(x + y) \times (x + z)$ $\qquad$ $(x \times y) + (x \times z)$ $\qquad$ distributive axioms
- $x + 0 = x$ $\qquad\qquad$ $x \times 1 = x$ $\qquad\qquad$ identity axioms
- $x + x' = 1$ $\qquad\qquad$ $x \times x' = 0$ $\qquad\qquad$ inverse axioms

The operations +, × and ' are called *addition, multiplication* and *complementation* respectively.

The word 'axiom' is used here rather than 'law', because these equations are part of the definition of a Boolean algebra, rather than statements to be proved.

The definition of a Boolean algebra is rather more abstract than any we have come across before, but it is precisely for that reason that it turns out to be so useful. It is well worth the effort to examine it a bit more closely.

First of all, the definition does not say what the elements of B are. They could be logical expressions, sets, or possibly something else again. When we are studying Boolean algebra, we do not need to know anything about the elements of B. When the time comes to apply the results of our study, we then *interpret* the elements of B as logical expressions, sets, or whatever.

A similar comment applies to the operations. We call them 'addition', 'multiplication' and 'complementation' (and denote them by familiar-looking symbols) simply because we have to call them *something*. It must be stressed that they are *not* addition, multiplication and complementation in the usual sense; the names and symbols are quite arbitrary. Likewise, 0 and 1 are not the numbers zero and one; they are arbitrary symbols for two particular elements of B.

Now suppose we interpret B to mean the set $\{T, F\}$, the operations +, × and ' to mean the logical connectives ∨, ∧ and ¬ respectively, the elements 0 and 1 to mean F and T respectively, and = to mean logical equivalence (denoted by ≡). We can rewrite the axioms in this interpretation; for example, the first commutative axiom becomes:

$$x \lor y \equiv y \lor x$$

---

[1]Boolean algebras are named after the English mathematician George Boole (1815–1864), who published a treatise on logical reasoning entitled *An Investigation of the Laws of Thought* in 1854.

This is a valid law of logic, and so are the other nine axioms when they are interpreted in this way. We conclude that propositional logic is a Boolean algebra.

Alternatively, we can interpret B to mean a power set $\mathcal{P}(A)$, and interpret $+$, $\times$, $'$, 0 and 1 as $\cup$, $\cap$, $^-$, $\varnothing$ and A respectively. We can then verify that the axioms of Boolean algebra are satisfied with this interpretation also.

There are other examples of Boolean algebras. For instance, if B is the set of all bytes (strings of 8 bits), with $+$, $\times$ and $'$ interpreted to mean the bitwise **or**, **and** and **not** operations, and 0 and 1 interpreted to mean the bytes 00000000 and 11111111 respectively, then the axioms of a Boolean algebra are satisfied.

You may be wondering why only some of the laws of logic (or laws of sets) have a corresponding axiom in Boolean algebra. The answer is that the other laws do not need to be included in the definition, because they can be deduced from the ten axioms.

EXAMPLE 8.1.1

Prove that the idempotent law for addition is true in any Boolean algebra:

$$x + x = x$$

SOLUTION

$x + x = (x + x) \times 1$	second identity axiom
$= (x + x) \times (x + x')$	first inverse axiom
$= x + (x \times x')$	first distributive axiom
$= x + 0$	second inverse axiom
$= x$	first identity axiom

It takes a little while to get used to seeing things like '$x + x = x$', which of course can never be a true statement in ordinary algebra. Just keep in mind that the symbols $+$, $\times$, 0 and 1 do not have their 'usual' meanings, and make sure that each step in the reasoning can be justified by one of the axioms (or by a law that has already been proved). Take particular care that you do not try to use a law of ordinary algebra that does not hold in Boolean algebra. For example, there are no 'cancellation laws' in Boolean algebra; if you have obtained the equation $x + y = x + z$, you are not permitted to cancel x from both sides.

With Example 8.1.1, you should be able to begin to see the value of studying Boolean algebra. Now that we have proved the additive idempotent law in Boolean algebra, it follows *automatically* that this law holds when we interpret it in logic, in power sets, or in any other system that satisfies the axioms. There is no need to prove it separately for the different interpretations.

There is a useful principle in Boolean algebra called the *duality principle*. The duality principle states that if we have proved a law of Boolean algebra, then the *dual* of that law is also true, where the dual is obtained by replacing + with ×, × with +, 1 with 0 and 0 with 1. The principle arises from the fact that the axioms occur in pairs, in which each axiom in a pair is the dual of the other one.

For example, the dual of the additive idempotent law we have just proved is the multiplicative idempotent law:

$$x \times x = x$$

There is no need to prove this law, since it follows from Example 8.1.1 and the duality principle. Alternatively, a proof of the multiplicative idempotent law can easily be constructed by replacing each line of the proof in Example 8.1.1 by its dual. (You are asked to do this in Exercise 4 at the end of the chapter.)

For convenient reference, here is a list of the main laws in Boolean algebra:

- $x'' = x$ double complement law
- $x + x = x$ $x \times x = x$ idempotent laws
- $(x + y)' = x' \times y'$ $(x \times y)' = x' + y'$ de Morgan's laws
- $x + 1 = 1$ $x \times 0 = 0$ annihilation laws
- $x + (x \times y) = x$ $x \times (x + y) = x$ absorption laws
- $0' = 1$ $1' = 0$ complement laws

These laws can all be proved from the axioms, just as we have done for the idempotent laws, although the proofs can be tricky unless you are given some hints. (De Morgan's laws are particularly difficult to prove.) You are guided through the proofs of some of the laws in the exercises at the end of the chapter.

For the rest of this chapter, there will no longer be any need to distinguish between the axioms and the laws derived from them, and so we will refer to all of them as 'laws'.

8.2 *Simplifying Boolean expressions*

We turn now to the problem of simplifying Boolean expressions. A *Boolean expression* is an expression involving one or more Boolean variables (denoted by letters such as x and y), together with the Boolean operations +, × and ', and the Boolean constants 0 and 1. We will see in Section 8.3 how the techniques for simplifying Boolean expressions can be applied to the practical problem of designing digital circuits.

A Boolean expression can be simplified by applying the laws of Boolean algebra until the simplest possible expression is obtained. Here are some examples showing how this is done.

EXAMPLE 8.2.1

| Simplify the Boolean expression $x \times (x \times y')'$.

SOLUTION

The presence of $(x \times y')'$ in the expression suggests that we should begin by applying the second De Morgan's law:

$$x \times (x \times y')' = x \times (x' + y'') = x \times (x' + y)$$

where the second equality is obtained using the double complement law.

The right-hand side is now in a form where the second distributive law can be used:

$$x \times (x' + y) = (x \times x') + (x \times y)$$

The expression $x \times x'$ can be simplified using the multiplicative inverse law:

$$(x \times x') + (x \times y) = 0 + (x \times y)$$

The expression on the right-hand side is very close to being in a form where we can use the additive identity law, but we need to use the additive commutative law first:

$$0 + (x \times y) = (x \times y) + 0 = x \times y$$

The last expression cannot be simplified further. The simplification of the original expression can therefore be written:

$$x \times (x \times y')' = x \times y$$

In Example 8.2.1, it was not too difficult to see which law to apply at each step. The next example is a little less straightforward.

EXAMPLE 8.2.2

Simplify the Boolean expression $[(x \times y) + (x' \times z)] + (y \times z')$.

SOLUTION

We notice that the terms $x \times y$ and $y \times z'$ have a common factor of y. This suggests that we should bring them together using the commutative and associative laws for addition, and then use the second distributive law to take out the common factor:

$$\begin{aligned}
[(x \times y) + (x' \times z)] + (y \times z') &= [(x' \times z) + (x \times y)] + (y \times z') \\
&= (x' \times z) + [(x \times y) + (y \times z')] \\
&= (x' \times z) + [(y \times x) + (y \times z')] \\
&= (x' \times z) + [y \times (x + z')]
\end{aligned}$$

We can now apply the first distributive law to this expression. At first sight, this seems to make things more complicated, but there is a chance that the terms containing x and z will combine together and simplify in some way. (If they don't, we will have to try something else.)

$$(x' \times z) + [y \times (x + z')] = [(x' \times z) + y] \times [(x' \times z) + (x + z')]$$

The factor $(x' \times z) + (x + z')$ can be tackled in more than one way. The simplest way is by using the second de Morgan's law:

$$\begin{aligned} (x' \times z) + (x + z') &= (x' \times z) + (x'' + z') \\ &= (x' \times z) + (x' \times z)' \\ &= 1 \end{aligned}$$

where we have used the additive inverse law to obtain the last line. Putting this result back in the earlier expression, we obtain:

$$(x' \times z) + [y \times (x + z')] = [(x' \times z) + y] \times 1 = (x' \times z) + y$$

The result of all these manipulations can be written as follows:

$$[(x \times y) + (x' \times z)] + (y \times z') = (x' \times z) + y$$

EXAMPLE 8.2.3

Interpret the result of Example 8.2.2 if x, y and z denote:

(a) logical expressions p, q and r;
(b) sets A, B and C.

SOLUTION

(a) $[(p \wedge q) \vee (\neg p \wedge r)] \vee (q \wedge \neg r) \equiv (\neg p \wedge r) \vee q$
(b) $[(A \cap B) \cup (\overline{A} \cap C)] \cup (B \cap \overline{C}) = (\overline{A} \cap C) \cup B$

8.3 *Digital circuits*

We saw in Chapter 2 that the circuitry in a binary digital computer operates with signals that can take only two values, which are usually represented by the symbols 0 and 1. We now want to take a closer look at how the design of binary digital circuitry is carried out. While we could use the notation of propositional logic for this purpose, it is more usual to carry out the necessary calculations in Boolean notation. This means that we will be working with the Boolean algebra for which the set B contains just the two elements 0 and 1 (corresponding to F and T respectively). The Boolean operations $+$, $\times$ and $'$ will represent respectively the logical connectives **or**, **and** and **not**. In short, we will simply be doing propositional logic in a different notation.

The truth tables for the operations (connectives), written in Boolean notation, are shown in Tables 8.1 to 8.3.

x	y	$x + y$
0	0	0
0	1	1
1	0	1
1	1	1

Table 8.1

x	y	$x \times y$
0	0	0
0	1	0
1	0	0
1	1	1

Table 8.2

x	x'
0	1
1	0

Table 8.3

Before we go any further, we will introduce some notational conventions that we will be using from now on. Firstly, we will omit the symbol $\times$, and simply write xy instead of $x \times y$. Secondly, we will omit the brackets in a sum or a product of three or more expressions, because according to the associative laws the result is the same whichever way such expressions are bracketed; for example, we will write $x + y + z$ and xyz rather than $(x + y) + z$ and $(xy)z$. Finally, we will assume that multiplication takes precedence over addition (just as it does in ordinary algebra), unless a different order has been specified using brackets; for example, the expression $x + yz$ will always mean $x + (yz)$ and never $(x + y)z$.

Using these conventions, the result of Example 8.2.2 can be written in the following way:

$$xy + x'z + yz' = x'z + y$$

We are now ready to look at digital circuits.

A *digital circuit* (or *logic gate circuit*) is an electronic device for performing a digital computation. It has a number of inputs, each of which is an electrical signal that can take one of two states, denoted by 0 and 1. For any given combination of inputs, the device computes one or more outputs, each of which is either 0 or 1.

In practice, a digital circuit is usually etched onto a microchip. We are not concerned here with the physical details of how the device is constructed, so it is convenient to think of a digital circuit as a 'box' with some input wires entering from the left and some output wires leaving from the right, as shown in Figure 8.1.

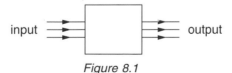

input ⇉ □ ⇉ output

Figure 8.1

A circuit of this type could be designed to perform any one of a wide variety of computational tasks. For example, we might want to design a circuit to add two numbers; the input would be the binary representations of the numbers, and the output would be the binary representation of their sum.

If a circuit has more than one output, we can think of it as consisting of several circuits, all with the same input, and each with a single output. For this reason, we will confine our attention to circuits with one or more inputs and one output.

It is useful to express these ideas in the mathematical language that was introduced in Chapters 5 and 6. Suppose we have a circuit with n inputs (and one output). We will assume that all possible combinations of n zeros and ones are valid inputs to the circuit. If B denotes the set $\{0, 1\}$, then the set of all possible inputs can be regarded as the Cartesian product B^n. Associated with each one of these inputs is an output of either 0 or 1. Mathematically, the circuit can be thought of as a function with domain B^n and codomain B:

$$f: B^n \to B$$

Functions of this type are called *Boolean functions*.

A Boolean function can be defined by listing all possible inputs together with their corresponding outputs. For example, we could define a Boolean function $f: B^2 \to B$ as shown in Table 8.4.

x	y	$f(x, y)$
0	0	1
0	1	0
1	0	1
1	1	1

Table 8.4

It is often simpler to define a Boolean function using a Boolean expression. For example, the Boolean function in the table above can be written as follows:

$$f: B^2 \to B, \quad f(x, y) = x + y'$$

We can verify that the two definitions of f are equivalent by constructing a table (Table 8.5), in which the third and fourth columns are obtained by referring to Tables 8.1 to 8.3.

x	y	y'	$x + y'$
0	0	1	1
0	1	0	0
1	0	1	1
1	1	0	1

Table 8.5

Table 8.5 is just a truth table written in Boolean notation.

Apart from having a convenient notation, there is another reason why we want to use Boolean expressions to represent digital circuits. A digital circuit can be constructed using devices known as *logic gates*. A logic gate is a simple digital circuit that corresponds to one of the logical connectives. By assembling logic gates, more complicated digital circuits can be built up. We will see shortly how the design of a digital circuit using logic gates corresponds to a Boolean expression.

Here are the definitions and symbols for the three logic gates we will be using in our work.

Definitions

An *OR gate* is a digital circuit with two inputs x and y, which produces an output of $x + y$. An OR gate is represented by the symbol shown in Figure 8.2(a).

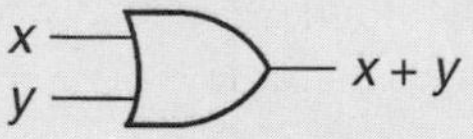

Figure 8.2(a)

An *AND gate* is a digital circuit with two inputs x and y, which produces an output of xy. An AND gate is represented by the symbol shown in Figure 8.2(b).

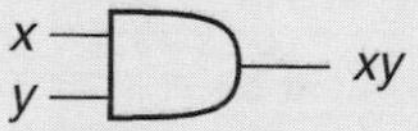

Figure 8.2(b)

An *inverter* (or *NOT gate*) is a digital circuit with one input x, which produces an output of x'. An inverter is represented by the symbol shown in Figure 8.2(c).

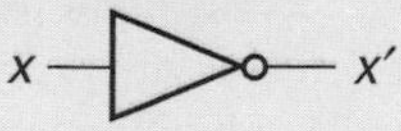

Figure 8.2(c)

These gates correspond to actual physical devices that can be assembled together into digital circuits. While the gates in most modern digital circuitry are microscopic in size and etched onto a microchip, it is also possible for gates to be constructed as components that can be assembled into a conventional electronic circuit. If 0 and 1 denote respectively the absence or presence of an electrical pulse, for example, then an OR gate is a device that produces an output pulse whenever a pulse is applied to at least one input, an AND gate produces an output pulse only if a pulse is applied to both inputs, and an inverter produces an output pulse whenever there is no pulse applied to the input.

At the moment, we are interested only with the logical behaviour of these gates, and so we need not concern ourselves here with details of their physical implementation. (For example, a real inverter would need a power supply in addition to the logical input, in order to supply the energy for the output pulse when no pulse is input.) Neither will we consider the problem of how to time the pulses so that they arrive at the gates at the moment they are needed; this is an important question as far as the actual construction of digital circuits is concerned, but it would take us too far afield to deal with it here.

If we have been given a Boolean function, corresponding to the inputs and outputs of a digital circuit, how can we design the circuit using only the three types of gates we have available? The answer is by finding a Boolean expression for the function, and drawing the circuit corresponding to that expression.

For example, consider the Boolean expression $xy' + y$. Starting with x and y as inputs, we can think of the expression as the result of a three-step process:

1. Evaluate y'.
2. Evaluate xy'.
3. Evaluate $xy' + y$.

Each step uses one of the Boolean operations, and corresponds to one of the logic gates. A circuit for evaluating the expression $xy' + y$ can be built up in this way using the gates, as shown in Figure 8.3.

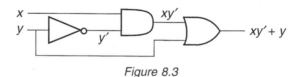

Figure 8.3

The output y' from the inverter is one of the inputs to the AND gate. The other input to the AND gate is x, so the output from the AND gate is xy'. This output in turn is one of the inputs to the OR gate, together with y. Notice that the wire from y is split, because y is needed as input to two of the gates.

If a Boolean expression contains any zeros or ones, they can easily be removed using the identity and annihilation laws. We can then draw the digital circuit for the resulting expression. (There is a trivial exception: the Boolean expressions 0 and 1 cannot be simplified further, and therefore have no corresponding circuits.)

EXAMPLE 8.3.1

Draw the digital circuit corresponding to the Boolean expression $(x' + y)'(x + y)$.

SOLUTION

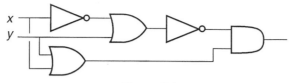

Figure 8.4

If two Boolean expressions are equal, then their corresponding circuits will behave in the same way. In practice, we would like our circuits to be as simple as possible, because simpler circuits tend to be faster and more reliable, and they occupy less space on a microchip. We can use Boolean algebra to replace a digital circuit with an equivalent one that is simpler.

EXAMPLE 8.3.2

Write down the Boolean expression corresponding to the digital circuit shown in Figure 8.5. Using the laws of Boolean algebra, obtain a simpler equivalent expression, and draw the corresponding circuit.

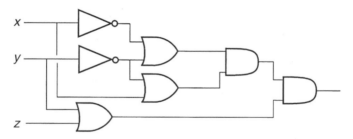

Figure 8.5

SOLUTION

The Boolean expression corresponding to this circuit is $(x' + y')(x + y')(y + z)$. It can be simplified in the following way:

$$\begin{aligned}
(x' + y')(x + y')(y + z) &= (x'x + x'y' + y'x + y'y')(y + z) \\
&= [0 + y'(x' + x) + y'](y + z) \\
&= (y'1 + y')(y + z) \\
&= (y' + y')(y + z) \\
&= y'(y + z) \\
&= y'y + y'z \\
&= 0 + y'z \\
&= y'z
\end{aligned}$$

(We have taken a few shortcuts here; some steps in the working involve the application of more than one law of Boolean algebra. You should check each step carefully to see that the laws have been followed.)

The simplified circuit is shown in Figure 8.6.

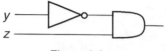

Figure 8.6

When we are working with more complicated digital circuits, it is convenient to allow the use of AND and OR gates with multiple inputs. These gates have the same symbols as the usual AND and OR gates, but are shown with varying numbers of input channels. For example, a three-input AND gate, corresponding to the expression *xyz*, is depicted as shown in Figure 8.7.

Figure 8.7

We will use multiple input gates in Example 8.4.4 in the next section.

8.4 *Disjunctive normal form and Karnaugh maps*

The method we used in the previous section for simplifying Boolean expressions relies on our being able to recognise at each step which of the laws of Boolean algebra can be applied to produce a simpler expression. Sometimes it might not be clear which law to use, or even whether the expression can be simplified at all. For these reasons, we will now look at a more systematic approach to the problem.

Suppose we want to find a Boolean expression corresponding to the function given in Table 8.6.

x	y	z	$f(x, y, z)$
0	0	0	1
0	0	1	0
0	1	0	0
0	1	1	1
1	0	0	0
1	0	1	1
1	1	0	0
1	1	1	0

Table 8.6

There is a purely mechanical way of obtaining a Boolean expression from a table like this, although it will usually not produce the simplest expression. Here is how it works.

For each line of the table for which $f(x, y, z) = 1$, we form a product of three factors, known as a *minterm*. Each factor in a minterm corresponds to one of the three variables x, y and z; the factor is the variable itself if the column for that variable contains 1, and the complement of the variable if the column contains 0.

In this example, we form the minterm for the first, fourth and sixth lines of the table, because these are the lines where $f(x, y, z) = 1$. The minterm for the first line is $x'y'z'$. The minterms for the fourth and sixth lines are $x'yz$ and $xy'z$ respectively.

An expression for $f(x, y, z)$ can now be written down; it is the sum of the minterms:

$$f(x, y, z) = x'y'z' + x'yz + xy'z$$

In the language of logic, $f(x, y, z)$ is true when x is false and y is false and z is false, or x is false and y is true and z is true, or x is true and y is false and z is true.

A Boolean function expressed in this way as a sum of distinct minterms is said to be written in *disjunctive normal form*. Every Boolean function can be written in disjunctive normal form in precisely one way (apart from trivial variations such as writing the minterms in a different order).

EXAMPLE 8.4.1

Let $f(x, y, z) = (xy')'(y + z')$. Write $f(x, y, z)$ in disjunctive normal form.

SOLUTION

The first step is to apply De Morgan's laws and the distributive laws to express $f(x, y, z)$ as a sum of terms:

$$(xy')'(y + z') = (x' + y)(y + z')$$
$$= x'y + x'z' + yy + yz'$$

Next, we deal with repeated variables within a term. Any term in which a variable or a complemented variable is repeated (such as xx or $x'x'$) can be simplified using the multiplicative idempotent law, and any term containing a variable and its complement (such as xx') can be eliminated using the multiplicative inverse law. For the present problem, we obtain:

$$x'y + x'z' + y + yz'$$

In order to put this expression into disjunctive normal form, we need to ensure that every term contains all three variables. This is done in the following way:

$$x'y + x'z' + y + yz'$$
$$= x'y(z + z') + x'z'(y + y') + y(x + x')(z + z') + yz'(x + x')$$
$$= x'yz + x'yz' + x'yz' + x'y'z' + xyz + xyz' + x'yz + x'yz' + xyz' + x'yz'$$

Now we use the additive idempotent law to eliminate the repeated terms. The result is:

$$f(x, y, z) = x'yz + x'yz' + x'y'z' + xyz + xyz'$$

There is no need to be concerned about the fact that the answer is more complicated than some of the expressions we obtained along the way. Disjunctive normal form is not meant to be a simplified form for a Boolean function, but rather a standard form in which all Boolean functions can be written.

Now that we know how to put an expression into disjunctive normal form, there remains the problem of simplifying it. A number of methods have been developed for doing this, but the use of Karnaugh maps, which we are about to describe, is probably the easiest for hand computation with expressions containing up to about four variables. If a larger number of variables is involved, there are other methods designed for implementation on a computer that are more suitable.

Suppose we have an expression written in disjunctive normal form and containing two variables x and y, for example $xy + x'y + x'y'$. We can construct a table for this expression with two rows, labelled x and x', and two columns, labelled y and y'. In the body of the table, we use a 1 to indicate that the minterm corresponding to that row and column is present, and leave the entry blank if the minterm is absent:

	y	y'
x	1	
x'	1	1

Notice that if there are two entries in the same row or column, then the corresponding minterms will have a common factor. In this example, the 1s in the second row correspond to the minterms $x'y$ and $x'y'$, which have a common factor of x', while the 1s in the first column correspond to minterms with a common factor of y. It is the presence of these common factors that allows us to simplify the expression. We identify the rows and columns where the common factors occur by drawing loops around the 1s:

A diagram constructed in this way is called a *Karnaugh map*.

The horizontal loop corresponds to $x'y + x'y'$. You can check using the laws of Boolean algebra that this expression can be simplified to x'. Similarly, the vertical loop corresponds to the terms $xy + x'y$, which can be simplified to y. There is no need to actually use the laws; we can read the answer directly off the Karnaugh map. The simplified expression is $x' + y$, where x' corresponds to the horizontal loop and y corresponds to the vertical loop.

If there had been no complete rows or columns of 1s (for example, if the 1 in the bottom left-hand corner had not been there), no simplification would have been possible. At the other extreme, if all the entries had been 1, the entire expression would have simplified to 1 (which we could have indicated by drawing a loop around all four 1s on the Karnaugh map).

Expressions containing only two variables are easy enough to simplify without using Karnaugh maps. The problems for which the method is most useful are those in which we need to simplify expressions containing three or four variables.

A Karnaugh map for an expression containing three variables x, y and z has two rows, labelled x and x', and four columns, labelled yz, $y'z$, $y'z'$ and yz'. In order for the method to work, it is essential that the column labels be written down from left to right in the order given. As before, a 1 means that the minterm is present in the expression, and a blank means it is absent.

There are several types of loops we can now draw on the Karnaugh map, corresponding to the ways in which the minterms can be combined and simplified:

- All eight 1s. (This applies only in the trivial case where all possible minterms are present.)
- Four 1s in a row.
- Four 1s in a square. This includes the 'broken' square consisting of the first and fourth columns, because we regard the right-hand edge as joining up with the left-hand edge. (It may be helpful to imagine that the Karnaugh map has been drawn around the curved surface of a cylinder.)
- Two adjacent 1s in a row or column, including 'broken' pairs such as the top left-hand and top right-hand corners.
- A single 1.

The larger the loop, the greater the resulting simplification. The general rule, therefore, is to begin by drawing as many of the loops containing four 1s as possible, followed by loops containing two 1s, and finishing with loops containing a single 1. Keep in mind that loops are allowed to overlap, and that every 1 must be included in at least one loop. Within these limits, the number of loops should be kept as small as possible.

EXAMPLE 8.4.2

Use a Karnaugh map to simplify the Boolean expression $xy'z + xyz' + xy'z' + x'y'z + x'y'z'$.

SOLUTION

We begin by setting up the table for the expression:

	yz	$y'z$	$y'z'$	yz'
x		1	1	1
x'		1	1	

There are no rows of four 1s, but there is a square of four 1s, so we put a loop around it. The remaining 1 cannot be included in a square of four 1s, but we can include it in a loop of two 1s. The Karnaugh map now looks like this:

	yz	$y'z$	$y'z'$	yz'
x		1	1	1
x'		1	1	

The square of four 1s yields the term y', because that is the only factor common to the four minterms. The loop of two 1s yields the term xz'. The simplified expression is therefore $y' + xz'$.

If a Boolean expression contains four variables, x, y, z and w, then the Karnaugh map has four rows and four columns. The rows are labelled (in order) $xy, x'y, x'y'$ and xy', and the columns are labelled $zw, z'w, z'w'$ and zw'. The following types of loops are available:

- All sixteen 1s (only if all possible minterms are present).
- Eight 1s in a four-by-two or two-by-four rectangle. This includes 'broken' rectangles such as the rectangle consisting of the first and fourth rows, because we identify the top and bottom edges with each other, and the left-hand and right-hand edges with each other.
- Four 1s in a row or column.

- Four 1s in a square, including 'broken' squares. In particular, this case includes the square comprising a 1 in each of the four corners of the table.
- Two adjacent 1s in a row or column, including 'broken' pairs.
- A single 1.

EXAMPLE 8.4.3

Use a Karnaugh map to simplify the Boolean expression $xyzw + xyz'w + xyzw' + x'yzw + x'yzw' + x'yz'w' + xy'zw + xy'z'w + xy'zw' + x'y'zw + x'y'zw'$.

SOLUTION

	zw	$z'w$	$z'w'$	zw'
xy	1	1		1
$x'y$	1		1	1
$x'y'$	1			1
xy'	1	1		1

The first and last columns together form a broken rectangle of eight 1s. We can pick up two of the three remaining 1s with a broken square of four 1s. The final 1 can be included in a loop of two 1s:

	zw	$z'w$	$z'w'$	zw'
xy	1	1		1
$x'y$	1		1	1
$x'y'$	1			1
xy'	1	1		1

The simplified expression is $z + xw + x'yw'$.

A variation of this method can be used if there are some inputs that we know will never arise in the application we have in mind, so that we don't care what output they produce. In problems of this type, the letter 'd' (for 'don't care') is used in the Karnaugh map to indicate that it does not matter whether the output corresponding to that particular set of inputs is 0 or 1. A 'd' does not have to be included in any loop, but it may be included in a loop if a simpler result can be obtained by doing so.

EXAMPLE 8.4.4

Design a digital circuit with four inputs, x, y, z and w, and with outputs according to Table 8.7.

x	y	z	w	Output
0	0	0	0	0
0	0	1	1	1
0	1	0	0	1
0	1	0	1	0
0	1	1	0	0
0	1	1	1	0
1	0	0	1	0
1	0	1	0	1
1	1	0	0	0
1	1	0	1	1
1	1	1	1	1

Table 8.7

Any combination of inputs not appearing in the table can be assumed never to occur.

SOLUTION

	zw	$z'w$	$z'w'$	zw'
xy	1	1		d
$x'y$			1	
$x'y'$	1	d		d
xy'	d		d	1

There are no rectangles of eight 1s. By using some of the 'd's, we can form two broken squares of four 1s. Another of the 1s can be looped into a pair, and the remaining 1 is in a loop of its own. Notice that two of the 'd's are not used.

	zw	$z'w$	$z'w'$	zw'
xy	1	1		d
$x'y$			1	
$x'y'$	1	d		d
xy'	d		d	1

The simplified expression is $y'z + xz + xyw + x'yz'w'$.

The digital circuit for an expression of this form is best handled using multiple input AND and OR gates as shown in Figure 8.8.

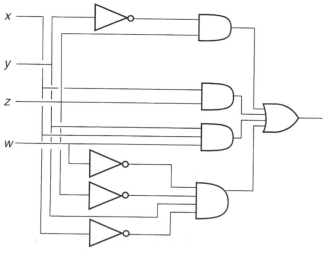

Figure 8.8

Before we conclude this example, it is worth noting one point. We drew the loops on the Karnaugh map by applying the rules exactly in the form we stated them, beginning with the larger loops. In fact, if we inspect the Karnaugh map, we can see that it could be simplified by removing the loop corresponding to the term xz. This example shows that the rules we have stated for drawing the loops on a Karnaugh map do not always produce the simplest Boolean expression.

Exercises

1. Explain why division is not a binary operation on **R**. Is division a binary operation on $\mathbf{R} - \{0\}$?

2. Determine which of the following are binary operations on the set indicated:
   (a) Addition on **N**
   (b) Subtraction on **N**
   (c) Subtraction on **J**
   (d) Multiplication on **Q**

3. For the Boolean algebra for which B is the set of all bytes, and $+$, $\times$ and $'$ denote bitwise **or, and** and **not**:
   (a) Verify the distributive axiom $x + (y \times z) = (x + y) \times (x + z)$ in the case $x = 10010100$, $y = 00111010$, $z = 01010001$.
   (b) Verify the inverse axiom $x + x' = 1$ in the case $x = 11001011$.

4. Construct a proof of the multiplicative idempotent law of Boolean algebra $x \times x = x$, by replacing each line of the proof of the additive idempotent law in Example 8.1.1 by its dual.

5. Use the axioms of Boolean algebra to prove the complement law $0' = 1$. (The first step uses one of the identity axioms.)

6. Use the axioms of Boolean algebra together with one of the idempotent laws to prove the annihilation law $x + 1 = 1$. (The first step uses one of the inverse axioms.)

7. Use the axioms of Boolean algebra together with one of the annihilation laws to prove the absorption law $x + (x \times y) = x$.

8. Prove the double complement law $x'' = x$, by proving first that $x \times x'' = x''$ and then that $x \times x'' = x$. (Begin both parts of the proof with the additive inverse axiom, replacing x by x' for the second part of the proof.)

9. In ordinary algebra, the *null factor law* for multiplication states: if $xy = 0$ then $x = 0$ or $y = 0$. Is the null factor law true:

   (a) in the logic interpretation of Boolean algebra;
   (b) in the power set interpretation of Boolean algebra;
   (c) in Boolean algebra?

10. Use the laws of Boolean algebra to simplify the following Boolean expressions:

    (a) $y \times [x + (x' \times y)]$
    (b) $[(x \times y') + x']'$
    (c) $[x + (y \times z)] \times (x' + z)$
    (d) $(x \times y)' + (y \times z)$

11. Interpret the result of Exercise 10(a) if x and y denote:

    (a) logical expressions p and q;
    (b) sets A and B.

12. Simplify the following Boolean expressions, and draw the digital circuits for both the original and simplified expressions:

    (a) $x'y + x(x + y')$
    (b) $(x + xy)'y$
    (c) $xz' + x'y + (yz)'$

13. Write down the Boolean expressions corresponding to the digital circuits shown in Figures 8.9(a) aned 8.9(b). For each expression, use the laws of Boolean algebra to obtain a simpler equivalent expression, and draw the corresponding circuit.

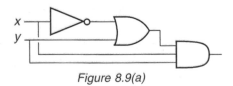

Figure 8.9(a)

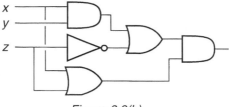

Figure 8.9(b)

14. Write the Boolean function given by Table 8.8 in disjunctive normal form:

x	y	z	$f(x, y, z)$
0	0	0	0
0	0	1	1
0	1	0	0
0	1	1	1
1	0	0	1
1	0	1	1
1	1	0	0
1	1	1	0

Table 8.8

15. Write the Boolean expression $(x' + z)(yz)'$ in disjunctive normal form.

16. Use Karnaugh maps to simplify the following Boolean expressions, converting them to disjunctive normal form first where necessary:

(a) $x'y + xy' + x'y'$
(b) $xy'z' + x'yz + x'y'z + x'yz' + x'y'z'$
(c) $xyz + xyz' + x'yz' + x'y'z'$
(d) $(x + y'z)(y + z')$
(e) $xyzw' + x'yzw + x'yz'w + x'yzw' + x'yz'w' + xy'z'w + xy'zw' + x'y'zw + x'y'z'w + x'y'zw'$
(f) $xyzw' + xyz'w' + x'yz'w + xy'zw' + x'y'zw + x'y'z'w'$
(g) $x(y + z'w)'$

17. In some problems involving Boolean expressions, a simpler answer can be obtained using the Karnaugh map of the complement of the expression to be simplified (i.e. interchange 1s and blanks in the Karnaugh map). The answer is then expressed as the complement of the simplified expression obtained from the Karnaugh map.

Use this method to simplify the following Boolean expressions:

(a) $xy'z' + x'y'z$
(b) $xyzw + xy'z'w + xy'z'w'$

18. Design a digital circuit with four inputs, x, y, z and w, and with outputs according to Table 8.9.

x	y	z	w	Output
0	0	0	1	1
0	0	1	0	0
0	1	0	0	1
0	1	1	1	0
1	0	0	1	0
1	1	0	0	0
1	1	0	1	0
1	1	1	0	0
1	1	1	1	1

Table 8.9

(Any combination of inputs that does not appear in the table can be assumed never to occur.)

19. The digits on the display of a certain calculator are each composed of seven light-emitting diodes (LEDs) arranged in the following layout:

The designs of the digits from 0 to 9 are shown below:

A digit to be displayed is input in its 4-bit binary representation to seven digital circuits, corresponding to the seven LEDs. The output of each circuit is 0 if the corresponding LED is off for that digit, and 1 if it is on.

For each LED, use a Karnaugh map (with 'don't care' entries where appropriate) to obtain a simple Boolean expression for the corresponding digital circuit, and draw a diagram for that circuit.

20. A digital circuit for adding 2 binary digits has two inputs: the 2 digits to be added; and two outputs: the units bit and the twos bit of the answer. (A circuit of this kind is called a *half adder*.)

By treating this as two circuits, each with a single output, use Karnaugh maps to obtain a Boolean expression for each circuit, and draw the corresponding circuit diagrams.

21. Consider a digital circuit for computing a column sum as part of the process of adding two binary numbers. A circuit of this kind (called a *full adder*) has three inputs: the 2 bits in the column to be added, and a carry bit (1 if there was 'one to carry' when the column immediately to the right was added, 0 otherwise). The circuit has two outputs: the 'sum' bit and the 'carry' bit.

By treating this as two circuits, each with a single output, use Karnaugh maps to obtain a Boolean expression for each circuit, and draw the corresponding circuit diagrams.

22. Design digital circuits for adding:

(a) two 2-digit binary numbers, using one half adder and one full adder;
(b) two 3-digit binary numbers, using one half adder and two full adders;

using the symbols shown below for a half adder and a full adder:

23. The behaviour of a NAND gate is defined by Table 8.10.

x	y	Output
0	0	1
0	1	1
1	0	1
1	1	0

Table 8.10

Its symbol is shown in Figure 8.10.

Figure 8.10

Using the results of Exercise 17 of Chapter 4, design digital circuits that are equivalent to the three gates AND, OR and NOT, using only NAND gates.

9. COMBINATORICS

9.1 Combinatorics and computing

Combinatorics is the branch of mathematics devoted to calculating the number of ways in which a specified process can be carried out. Problems of this type often occur in computing. A typical example is the problem of determining how many times a particular sequence of steps in an algorithm will be executed, as part of the larger problem of estimating how long the algorithm will take to run. For example, a simple algorithm for determining the shortest path for a signal to travel through a communications network might begin by calculating the lengths of all the possible paths through the network. A better algorithm might arrive at the correct answer after calculating just the lengths of a particular selection of the paths. In order to determine how efficient or inefficient these algorithms are, and to be able to compare them with each other and with other algorithms, we must first answer the question: 'How many such paths are there altogether?'

In this chapter, we present some basic counting techniques that can be used to solve problems of this type.

9.2 The Multiplication principle

Suppose you have a printer that can print in five different typefaces: Pica, Élite, Courier, Times Roman and Sans Serif. Suppose also that each typeface can appear in one of three styles: normal, bold and italic. A font is a combination of a particular typeface with a particular style, such as Courier Italic. How many fonts are there altogether?

The question can be answered very simply. Since there are five typefaces, each of which can be paired with one of three styles, the total number of fonts is $5 \times 3 = 15$. The different possible combinations are illustrated in the tree diagram in Figure 9.1.

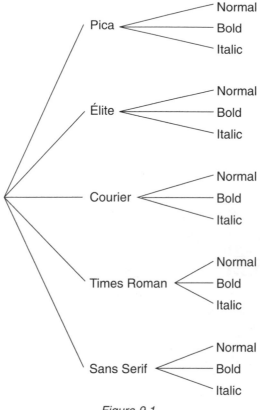

Figure 9.1

The problem can also be expressed in the notation of sets. Let *T* be the set of typefaces, and let *S* be the set of styles. A font such as Courier Italic can be thought of as an ordered pair — (Courier, Italic). The set of all such ordered pairs is the Cartesian product $T \times S$, and the problem asks for the cardinality of $T \times S$. We obtained the answer by multiplying the cardinality of *T* by the cardinality of *S*.

This example illustrates a basic principle in combinatorics known as the *Multiplication principle*. The Multiplication principle states that the cardinality of the Cartesian product of two finite sets *X* and *Y* equals the cardinality of *X* multiplied by the cardinality of *Y*. We can write this result in symbols as follows:

$$|X \times Y| = |X| \times |Y|$$

More generally, we can obtain the cardinality of a Cartesian product $X_1 \times X_2 \times \ldots \times X_n$ of a number of finite sets by multiplying together the cardinalities of the sets $X_1, X_2, \ldots, X_n$. The Multiplication principle is usually stated in this more general form:

$$|X_1 \times X_2 \times \ldots \times X_n| = |X_1| \times |X_2| \times \ldots \times |X_n|$$

For many practical applications, a more intuitive (though somewhat wordier) version of the Multiplication principle is useful:

If a selection process consists of n steps, where the selection in the first step can be done in k_1 ways, the selection in the second step can be done in k_2 ways, and so on, then the total number of possible selections is $k_1 k_2 \ldots k_n$.

We have actually met the Multiplication principle already. In Chapter 3, we used it to calculate the number of distinct integers that could be stored in 16 bits, by arguing that there were two possibilities for the first bit, two possibilities for the second bit, and so on, giving an answer of 2^n integers altogether. We also used the Multiplication principle in Chapter 5, when we showed that any set with n elements has 2^n subsets.

EXAMPLE 9.2.1

The usercodes on a certain computer consist of 3 letters, followed by 3 digits, followed by a letter, for example XYZ123A. (Assume that no distinction is made between upper-case and lower-case letters.)

(a) How many different usercodes can be constructed altogether?
(b) In how many of these usercodes does the digit 0 occur at least once?

SOLUTION

(a) The first character may be chosen to be any one of the 26 letters of the alphabet, and so may the second and third characters. The fourth, fifth and sixth characters are each chosen from the 10 digits, and the seventh character is chosen from the 26 letters. By the Multiplication principle, the total number of different usercodes is:

$$26 \times 26 \times 26 \times 10 \times 10 \times 10 \times 26 = 456\,976\,000$$

(b) Rather than try to count directly the number of usercodes containing at least one zero, it is simpler to count the number of usercodes that contain no zeros, and to subtract the answer from the result of part (a). If no zeros are allowed, then there are only nine choices available for each digit rather than ten, so the number of usercodes with no zeros is:

$$26 \times 26 \times 26 \times 9 \times 9 \times 9 \times 26 = 333\,135\,504$$

Therefore the number of usercodes containing at least one zero is:

$$456976000 - 333135504 = 123\,840\,496$$

9.3 *Permutations*

A common task in computing is that of sorting a list of items into order according to some criterion, such as alphabetical or numerical order. We will look at an example of a sorting algorithm in Chapter 13. In the meantime, a question that comes to mind in this context is: In how many different orders can a set of items be listed?

If there are three items, for example, say *a*, *b* and *c*, then a bit of trial and error reveals that there are six different ways in which they can be arranged:

$$abc, acb, bac, bca, cab, cba$$

An arrangement of the elements of a set into a particular order is called a *permutation* of that set. We have just listed the six permutations of the set $\{a, b, c\}$.

How many permutations are there of a set with *n* elements? This question can be answered by applying the Multiplication principle in the following way. The first element of the permutation can be any one of the *n* elements. Once this element has been chosen, the second element can be any one of the remaining $n - 1$ elements. Continuing in this way, the third element can be any one of the remaining $n - 2$ elements, the fourth element can be any one of the remaining $n - 3$ elements, and so on. There is of course only one possible choice for the last element. According to the Multiplication principle, the total number of permutations is:

$$n(n - 1)(n - 2) \ldots \times 3 \times 2 \times 1$$

We recognise this from our earlier work as *n* factorial (*n*!). The result can be stated as follows: *The number of permutations of a set with n elements is n!*.

EXAMPLE 9.3.1

How many anagrams (permutations of the letters) are there of the following words?

(a) ANSWER   (b) PERMUTE   (c) LITTLE

SOLUTION

(a) Since ANSWER has six different letters, the number of anagrams is $6! = 720$.

(b) There are 7! ways of arranging seven letters; however, this assumes that all the letters are different. In the present case, the number of anagrams is only half that number, because interchanging the two Es in an anagram does not produce a new anagram. The answer is $\dfrac{7!}{2} = 2520$.

(c) There are 6! permutations of six letters, but again we need to take care to avoid double counting. In each of these 6! arrangements, we can swap the Ts or the Ls or both without obtaining new anagrams. The number of different anagrams is therefore $\frac{6!}{4} = 180$.

Note that in each case the original word is counted as one of the anagrams.

A more general permutation problem arises if the elements to be arranged are chosen from a larger set. Suppose we want to select r elements from a set with n elements (where $n > r$), and arrange those r elements in a particular order. In how many ways can this be done?

This question can be answered in a similar manner to the previous one. The first element of the permutation can be any one of the n elements in the larger set. The second element must then be chosen from the remaining $n - 1$ elements, the third element from the remaining $n - 2$ elements, and so on, until r elements have been chosen. The last (rth) element is chosen from the $n - r + 1$ elements remaining at that stage. By the Multiplication principle, the total number of ways this can be done is:

$$n(n - 1)(n - 2) \ldots (n - r + 1)$$

This expression is denoted by ^nP_r. It is not difficult to see that we can also write ^nP_r in terms of factorials in the following way:

$$^nP_r = \frac{n!}{(n - r)!}$$

EXAMPLE 9.3.2

As part of a market research survey, you are shown a list of 20 types of chocolate bar, and asked to list your five favourite bars in order of preference. How many different responses to this question are there?

SOLUTION

You are selecting five bars from a larger set of 20 bars. Since the order in which the five bars are listed is important (i.e. the same selection in a different order counts as a different response), this is a problem involving permutations. The answer is:

$$^{20}P_5 = 20 \times 19 \times 18 \times 17 \times 16 = 1\,860\,480$$

9.4 *Combinations*

In Section 9.3, we investigated the problem of selecting a certain number of items from a set, where the order in which the items are selected is taken into account. A more common problem in practice is to ascertain the number of ways in which a selection can be made, without distinguishing between different orders. Problems of this kind are called *combinations* problems.

For example, suppose that the rules of a lottery specify that entrants must select eight different numbers chosen from the natural numbers from 1 to 45 inclusive. In how many different ways can this selection be made?

As a first step towards solving the problem, we can calculate the number of choices if order *is* taken into account; it is $^{45}P_8$. This number is not the answer to the problem, of course, because we have counted the same selection many times, once for each different way in which the eight numbers can be arranged. In fact, we can state precisely how much too large it is; since there are 8! ways of arranging the eight chosen numbers, $^{45}P_8$ is too large by a factor of 8!. The correct answer to the problem is therefore $\dfrac{^{45}P_8}{8!}$, which equals 215553195.

This example gives us a clue to the solution of the combinations problem in general. Suppose we want to select r items without replacement from a set of n items, without taking into account the order in which the selection is made. (In the language of sets, we want to choose a subset with r elements from a set with n elements.) In how many ways can this be done?

The number of selections with order taken into account is ^nP_r. Now, if order is *not* to be taken into account, then this number is too large by a factor of $r!$, because $r!$ is the number of permutations of the r selected elements. The correct answer is therefore $\dfrac{^nP_r}{r!}$. Using the fact that $^nP_r = \dfrac{n!}{(n-r)!}$, we obtain the following expression as the answer:

$$\frac{n!}{r!(n-r)!}$$

This expression is denoted by $\dbinom{n}{r}$ or ^nC_r. It is the number of ways that r items can be chosen from a set of n items. Equivalently, it is the number of r-element subsets of an n-element set. Given a particular value of n, the values of $\dbinom{n}{r}$ (as r ranges from 0 to n) are referred to as the nth *binomial coefficients*, because of their occurrence in the binomial theorem in algebra.

Using the alternative expression for ^nP_r, we obtain another expression for $\dbinom{n}{r}$ that is generally more useful for computations:

$$\binom{n}{r} = \frac{n(n-1)(n-2)\ldots(n-r+1)}{r!}$$

EXAMPLE 9.4.1

A test is to be generated by selecting six questions from a test bank consisting of 50 questions. How many different tests are possible, assuming that the order in which the questions appear on the test is not taken into account?

SOLUTION

The number of different tests is:

$$\binom{50}{6} = \frac{50 \times 49 \times 48 \times 47 \times 46 \times 45}{6 \times 5 \times 4 \times 3 \times 2 \times 1} = 15\ 890\ 700$$

A comment about evaluating these kinds of expressions on a calculator is appropriate here. Some calculators have keys for evaluating ^nP_r and $\binom{n}{r}$ directly. If a calculator with these facilities is not available, it may be necessary first to perform some simple cancellations by hand, in order to avoid overflow problems on the calculator. (In Example 9.4.1, the 48 in the numerator cancels with the factors 6, 4 and 2 in the denominator, and the 45 in the numerator can be divided by the 5 and the 3 in the denominator, leaving the expression $50 \times 49 \times 47 \times 46 \times 3$ to be evaluated on a calculator.)

EXAMPLE 9.4.2

A task group consisting of three women and three men is to be chosen from the members of an organisation consisting of 23 women and 18 men. In how many ways can this be done?

SOLUTION

The three women can be chosen in $\binom{23}{3}$ ways. For each of these choices, the three men can be chosen in $\binom{18}{3}$ ways. By the Multiplication principle, the total number of ways is $\binom{23}{3}\binom{18}{3} = \frac{23 \times 22 \times 21}{3 \times 2 \times 1} \times \frac{18 \times 17 \times 16}{3 \times 2 \times 1} = 1\ 445\ 136$ ways.

In some problems, combinations appear in a rather more subtle way, as the next example illustrates.

EXAMPLE 9.4.3

A motorist wants to travel from A to B using the grid of streets shown in Figure 9.2. Assuming that no 'doubling back' is allowed at any stage of the journey, in how many ways can this be done?

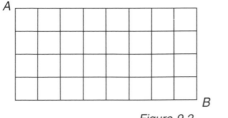

Figure 9.2

SOLUTION

Since there is no doubling back, a journey from A to B consists of a list of segments, where each segment is the length of a single block, either south (S) or east (E). For example, one such journey is ESSEEESESEEE. There must be 8 segments east and 4 segments south, and these 12 segments can be arranged in any order. The problem then becomes: in how many different ways can 8 Es and 4 Ss be arranged?

This can actually be treated as a combinations problem, although at first sight it might not appear to be one. The number of such arrangements is just the number of different ways of selecting the positions of the 4 Ss from the 12 available positions in the list. (Once the positions of the Ss have been chosen, the Es must occupy all the other positions.) The answer is:

$$\binom{12}{4} = \frac{12 \times 11 \times 10 \times 9}{4 \times 3 \times 2 \times 1} = 495$$

There are 495 different journeys.

The binomial coefficients $\binom{n}{r}$ satisfy a number of identities. One of the most useful of these is the following identity:

$$\binom{n}{r} = \binom{n}{n-r}$$

This result is easily proved by writing out each side in terms of factorials. It is useful in computations because it allows a binomial coefficient with $r > \dfrac{n}{2}$ to be replaced by one with $r < \dfrac{n}{2}$.

For example, rather than evaluate $\binom{30}{25}$ directly, it is simpler to replace it with $\binom{30}{5}$ using the identity above:

$$\binom{30}{25} = \binom{30}{5} = \frac{30 \times 29 \times 28 \times 27 \times 26}{5 \times 4 \times 3 \times 2 \times 1} = 142506$$

Some other useful identities are:

$$\binom{n}{0} = 1, \ \binom{n}{n} = 1$$

$$\binom{n}{r} = \binom{n-1}{r-1} + \binom{n-1}{r} \quad (n \geq 1)$$

The first two identities follow immediately from the definition of $\binom{n}{r}$ (remembering that $0! = 1$). You are asked to prove the third identity in the exercises.

The third identity forms the basis of a simple method for generating and displaying the binomial coefficients row by row. The top row of the display contains just the number 1, which is the value of $\binom{0}{0}$. Beneath it are the values of $\binom{1}{0}$ and $\binom{1}{1}$, the binomial coefficients corresponding to $n = 1$. The next row contains the values of $\binom{2}{0}$, $\binom{2}{1}$ and $\binom{2}{2}$, the binomial coefficients corresponding to $n = 2$, and so on. The first and last entries in each row are 1, because $\binom{n}{0} = \binom{n}{n} = 1$. Each other entry is obtained by adding the two numbers immediately above it, because $\binom{n}{r} = \binom{n-1}{r-1} + \binom{n-1}{r}$. The first six rows (corresponding to values of n from 0 to 5) are shown below:

```
              1
            1   1
          1   2   1
        1   3   3   1
      1   4   6   4   1
    1   5  10  10   5   1
```

This way of displaying the binomial coefficients is called *Pascal's triangle*, after the French mathematician Blaise Pascal (1623–1662) who investigated its properties, although it had already been known to earlier mathematicians.

> ## Exercises

1. Suppose you are offered a choice of tea or coffee, with full milk, reduced milk or no milk, and either with or without sugar. Express all the possible drinks you could have as a Cartesian product of sets, and depict them on a tree diagram.

2. How many different car number plates consisting of 3 letters followed by 3 digits are there altogether?

3. The telephone numbers in a certain area consist of 7 digits. The first digit of a telephone number is not permitted to be 0 or 1.

   (a) How many such telephone numbers are there altogether?
   (b) How many of the telephone numbers do not contain any zeros?
   (c) How many of the telephone numbers contain at least one zero?

4. Australian postcodes consist of 4 digits, but only the digits 0, 2, 3, 4, 5, 6 and 7 are used for the first digit, and if the first digit is 0 then the second digit is always 8. How many different postcodes are possible with these restrictions?

5. The DX code on photographic film canisters (which is used by some cameras to set the film speed and number of exposures automatically) consists of 12 regions, numbered from 1 to 12. Each region is either conducting or non-conducting.

   (a) If all possible combinations of 'conducting' and 'non-conducting' can occur, how many different codes are possible?
   (b) In fact, regions 1 and 7 are always conducting. In addition, at least one of regions 5 and 6 is always conducting. How many codes are possible with these restrictions?

6. On a certain computer system, a valid password consists of exactly 8 characters. The first character must be a letter, and the remaining characters must be letters or digits. (Assume that no distinction is made between upper-case and lower-case letters.) In order to be valid, a password must contain at least one digit. How many valid passwords are there?

7. A combination lock for a bicycle has four dials, each with the numbers from 1 to 6. If you have forgotten the combination that opens the lock, how long will it take you to try all the combinations, at a rate of one every 5 seconds?

8. A program to generate all the permutations of a set is run on a computer that writes the output to a file at a rate of 1000 permutations per second. How long will it take the computer to generate all the permutations of a set with:

   (a) 10 elements?
   (b) 15 elements?

9. Evaluate:

    (a) $^{10}P_6$

    (b) $^{12}P_8$

    (c) $^{13}C_5$

    (d) $\binom{15}{11}$

10. In how many ways can a club with 45 members select a president, a vice president, a secretary and a treasurer, if no member is permitted to hold more than one office?

11. The test marks for a class of 12 students are as follows:

$$15, 13, 18, 15, 7, 12, 10, 13, 9, 5, 15, 17$$

These marks may be entered into a statistical analysis program in any order. How many different orders are there?

12. The Information Systems department at a university has 16 members of staff. The department has been asked to choose five of its members to serve on a faculty committee.

    (a) In how many ways can the members who are to serve on the committee be chosen?

    (b) If seven of the 16 members of the department are female, how many five-member committees can be formed in which exactly two of the members are female?

13. A test bank of questions is to be set up so that tests consisting of ten questions can be generated by selecting questions from the test bank. What is the minimum number of questions needed in the test bank in order to ensure that at least 1000 different tests can be generated? (Assume that the order in which the questions appear on a test is not important.)

14. The position of a robotic arm above a printed circuit board is described by a pair of co-ordinates, giving the distances of the point from the left-hand edge and the lower edge of the board, in that order. The arm is initially at the bottom left-hand corner of the board, which has co-ordinates $(0, 0)$. Two other points on the board are A, with co-ordinates $(6, 3)$, and B, with co-ordinates $(3, 5)$. The arm can only move in unit steps parallel to the sides of the board.

    In how many different ways can the arm move:

    (a) from the initial position to A;

    (b) from the initial position to A, and then to B;

    (c) from the initial position to A, then to B, and back to the initial position;

    (d) from the initial position to A and B in either order, returning to the initial position?

    (Assume that the arm always takes one of the shortest paths from one point to another, with no doubling back.)

15. Prove the identity $\binom{n}{r} = \binom{n-1}{r-1} + \binom{n-1}{r}$.

16. Prove that $^{n}P_r = n \, ^{n-1}P_{r-1}$.

17. Write down the rows of Pascal's triangle corresponding to $n = 6$ and $n = 7$.

18. By interpreting $\binom{n}{r}$ as the number of r-element subsets of an n-element set, find the value of $\binom{n}{0} + \binom{n}{1} + \binom{n}{2} + \ldots + \binom{n}{n}$.

19. Without referring to $\binom{n}{r}$, explain why a set with n elements has:

    (a) n 1-element subsets;

    (b) n $(n-1)$-element subsets;

    (c) the same number of $(n-r)$-element subsets as r-element subsets.

10. INTRODUCTION TO GRAPH THEORY

10.1 *What is a graph?*

The objects that we study in the branch of mathematics known as graph theory are not graphs drawn with x and y axes. In this chapter, the word 'graph' refers to a structure consisting of points (called 'vertices'), some of which may be joined to other vertices by lines (called 'edges') to form a network. Structures of this type abound in computing. The computers on a site may be connected into a local area network, which in turn may be linked to national and international communications networks. The circuitry inside a computer (which we represented schematically by digital circuit diagrams in Chapter 8) is another example of a graph or network structure. At a more abstract level, we saw in Chapter 5 how a relation on a set can be depicted using a diagram that takes the form of a graph.

There is a particular type of graph called a *tree*, which we will study in Chapter 11. Trees are used in computing to represent the structure of expressions; we saw an example of a logical expression tree in Chapter 4. Trees can also be used to represent decision processes, in which we are faced with two or more choices at each step of a procedure; the tree diagrams in Chapter 9 were of this type.

In this chapter, we will introduce graphs, study their basic properties, and investigate some practical problems in which they can be applied. In Chapter 11, we will study trees and investigate some of their applications to computing.

10.2 *Basic concepts in graph theory*

Definition

A *graph* consists of a non-empty set of points, called *vertices* (singular: *vertex*), and a set of lines or curves, called *edges*, such that every edge is attached at each end to a vertex.

An example of a graph is shown in Figure 10.1, with the edges and vertices labelled.

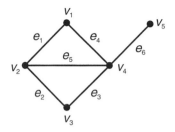

Figure 10.1

An edge is said to be *incident* to the vertices to which it is attached. For example, the edge e_1 is incident to the vertices v_1 and v_2 in the graph in Figure 10.1.

Two vertices that are joined by an edge are said to be *adjacent*. In the graph in Figure 10.1, v_1 and v_2 are adjacent, while v_1 and v_3 are not adjacent.

A graph such as the one shown in Figure 10.1 could arise in many situations of practical interest. For example, a graph could represent a road system, in which the edges are the roads and the vertices are the towns and road junctions. Alternatively, the vertices might represent computer laboratories in a local area network (LAN), with the edges representing the links in the network.

More abstract interpretations are also possible. For example, the vertices could be sporting teams, with an edge joining two vertices if those teams have played against each other during the current season.

Before we go any further, it is important that we clarify just what is and what is not allowed in a graph.

Firstly, a graph does not have to be in one piece. For example, Figure 10.2 shows a perfectly legitimate graph.

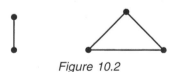

Figure 10.2

If a graph *is* in one piece, it is said to be *connected*. For some applications we will only be interested in connected graphs, but we do not make being connected part of the definition of a graph.

The second point to notice is that there is no restriction on the numbers of edges and vertices a graph may have (except that it must have at least one vertex). It is permissible for a graph to have no edges; such graphs are called *null* graphs. It is also permissible for the number of vertices or edges of a graph to be infinite; however, we will assume from now on that all the graphs we will be dealing with have a finite number of vertices and edges.

A graph may have *loops* (an edge from one vertex to the same vertex), as in Figure 10.3.

Figure 10.3

A graph may also have *parallel* edges (two or more edges linking the same pair of vertices), as in Figure 10.4.

Figure 10.4

In some applications of graphs, loops and parallel edges make sense. For example, if a graph represents a road system, there may be two towns with two or more roads joining them, and there may be a road that leaves a town and returns to it. There are other applications in which loops and multiple edges do not arise — a graph representing a local area network, for example.

A graph is said to be *simple* if it has no loops and no parallel edges.

Notice that we have not assigned a direction to the edges of a graph. In some applications it would make sense for each edge to have a direction. For example, in the technique known as critical path analysis, the vertices of a graph represent tasks to be carried out, and an edge from one vertex to another indicates that one task has to be completed before the other can begin. This situation can be depicted using directed edges (edges with arrows drawn on them), and a graph with directed edges is called a *directed* graph. The graphical depiction of a relation, which we met in Chapter 5, is an example of a directed graph. The graphs we will be studying in this chapter are all undirected graphs.

We now come to a subtle but important point about the way graphs are drawn. When we draw a graph, all that matters is which vertices are adjacent (joined by an edge) and which are not; the precise location of the vertices and the lengths and shapes of the edges are irrelevant. This means that there will be many different ways of drawing the same graph, some of which may look very different from others.

For example, the two diagrams in Figure 10.5 represent the *same* graph.

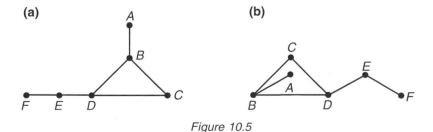

Figure 10.5

In order to make the relationship between these two graphs clearer, corresponding vertices have been labelled with the same letter. The graphs are essentially the same (they are *isomorphic*, to use the technical term), because the first graph can be 'rearranged' to look like the second. In this rearranging process, we are free to move the vertices around, and stretch and bend the edges, as much as we like. We may even draw two edges so that they cross at a point where there is no vertex. The only restrictions are that we may not create new edges or vertices, delete any edges or vertices, or reattach edges to different vertices.

Imagine that the graph in Figure 10.5(a) is made of buttons (the vertices) joined by pieces of elastic string (the edges). We may move the buttons and string around however we like, as long as we don't break any of the strings. In particular, we can rotate the first graph anticlockwise so that it looks like the one shown in Figure 10.6.

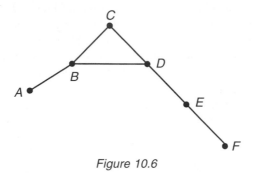

Figure 10.6

If we flip the edge *AB* over into the triangle *BCD*, and bend *DEF* up a bit, we now obtain the graph shown in Figure 10.5(b) without breaking or rejoining edges or creating new ones.

We will have more to say in Section 10.4 about graphs being isomorphic.

A useful way to get a feel for graphs is to look for all the simple graphs with a small number of vertices. We will do this now, by finding all the simple graphs with up to three vertices.

If a simple graph has one vertex, then it must have no edges, because a simple graph is not permitted to have loops. Therefore there is only one simple graph with one vertex; it looks like Figure 10.7(a).

Figure 10.7(a)

If a simple graph has two vertices, then either there is an edge joining the vertices or there is not. The two possibilities are shown in Figure 10.7(b).

●   ●   ●━━━━●

Figure 10.7(b)

With three vertices, things start to get more complicated. It doesn't matter whether we draw the vertices in a straight line or in a triangle, because we may move them around however we like and the graph will still be essentially the same graph. Suppose we arrange the vertices in a triangle. We find that there are four different simple graphs with three vertices (Figure 10.7(c)).

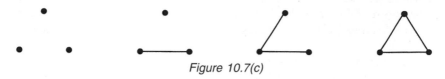

Figure 10.7(c)

These four graphs are all different (they must be, because they have different numbers of edges), and any simple graph with three vertices is isomorphic to one of these four. (For example, a graph with three vertices 'in line' can be obtained by 'straightening out' the third graph in Figure 10.7(c).)

You are asked in Exercise 2 to draw all 11 simple graphs with four vertices. This can be tricky, but it is a useful exercise for becoming familiar with graphs. (If you think you have found more than 11, you will need to look carefully to see whether you have included graphs that appear to be different but are actually isomorphic.)

There are 34 simple graphs with five vertices. The number of simple graphs increases rapidly as the number of vertices increases. There is a known method for calculating the number of simple graphs with n vertices, but as it is very complicated we will not give it here.

We now return to graphs in general (not necessarily simple graphs). In order to proceed further, we need the following definitions.

Definitions

> The *degree* of a vertex v in a graph is the number of edges incident to v, with a loop counting 2 towards the degree of the vertex to which it is incident. The degree of v is denoted by deg(v).
> A vertex with degree 0 is said to be *isolated*.

For example, the vertices of the graph shown in Figure 10.8 have the degrees indicated.

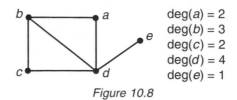

deg(a) = 2
deg(b) = 3
deg(c) = 2
deg(d) = 4
deg(e) = 1

Figure 10.8

There is a simple relationship between the degrees of the vertices of any graph and the number of edges of the graph. It is given by the following theorem.

Theorem

In any graph, the sum of the degrees of the vertices equals twice the number of edges.

If V and E denote respectively the set of vertices and the set of edges of a graph, then the theorem can be written in symbols as follows:

$$\sum_{v \in V} \deg(v) = 2|E|$$

The theorem follows immediately from the fact that if we add the degrees of all the vertices, every edge will be counted twice, since each edge is incident to two vertices.

EXAMPLE 10.2.1

Verify for the graph shown in Figure 10.8 that the sum of the degrees of the vertices equals twice the number of edges.

SOLUTION

$$\sum_{v \in V} \deg(v) = \deg(a) + \deg(b) + \deg(c) + \deg(d) + \deg(e)$$
$$= 2 + 3 + 2 + 4 + 1$$
$$= 12$$
$$= 2 \times 6$$
$$= 2|E|$$

A more surprising result is the following one, which is widely known by the curious name of the 'Handshaking lemma'. (A lemma is a result that is used primarily to prove other theorems rather than being of interest in its own right.)

'Handshaking lemma'

In any graph, the number of vertices with odd degree is even.

For example, the graph in Example 10.2.1 (Figure 10.8) has two vertices with odd degree (b and e), and two is an even number. In general, the Handshaking lemma asserts that if we count the number of vertices with odd degree in a graph, we must obtain an even number as the answer.

The name of the lemma arises from an amusing way of interpreting this result. Imagine a room full of people, some of whom shake hands with some of the other people in the room. The situation can be represented as a graph, in which the vertices represent the people in the room, and two vertices are adjacent if the corresponding people shake hands. Even if we do not know how many people there are, or how many times and with whom each person shakes hands, we can nevertheless make one assertion with confidence: *the number of people who shake hands an odd number of times is even.*

The 'Handshaking lemma' can be proved in the following way. We begin with the result of the previous theorem:

$$\sum_{v \in V} \deg(v) = 2|E|$$

The left-hand side of this equation can be split into two terms: the sum taken over the set V_{odd} of vertices with odd degree, and the sum taken over the set V_{even} of vertices with even degree:

$$\sum_{v \in V_{odd}} \deg(v) + \sum_{v \in V_{even}} \deg(v) = 2|E|$$

The second term on the left-hand side is a sum of even numbers, and is therefore equal to an even number. The right-hand side is clearly an even number. Thus the first term on the left-hand side is the difference between two even numbers, and so must be an even number. Now, this term is a sum of odd numbers; but an even result can be obtained by adding odd numbers together only if there is an even number of them. Hence the number of vertices with odd degree is even. This completes the proof.

Definition

A simple graph is *complete* if each vertex of the graph is adjacent to every other vertex.

A complete graph with five vertices is shown in Figure 10.9.

Figure 10.9

We can obtain a formula for the number of edges in a complete graph with n vertices using the following reasoning. Each edge in a complete graph corresponds to a selection of two distinct vertices from the set of n vertices, without taking order into account. This selection can be carried out in $\binom{n}{2}$ ways. Hence the number of edges is $\binom{n}{2}$, or $\dfrac{n(n-1)}{2}$.

Definition

The *complement* $\overline{G}$ of a simple graph G is the graph with the same vertices as G, such that any two vertices are adjacent in $\overline{G}$ if and only if they are not adjacent in G.

For example, if the vertices of G represent sporting teams, and an edge indicates that two teams have already played a match against each other this season, then $\overline{G}$ shows the matches that are still to be played if each team plays once against every other team during the season.

10.3 *The matrix representation of a graph*

While a pen-and-paper drawing of a graph is convenient for working with small graphs by hand, it is not suitable if a graph is to be handled on a computer. For this purpose, other ways of representing graphs have been developed. The most commonly used representation is known as an *adjacency matrix*.

If G is a graph with n vertices, labelled $v_1, v_2, \ldots, v_n$, then the adjacency matrix of G is an $n \times n$ matrix whose entries are given by the following rule: *The entry in the ith row and the jth column is the number of edges from v_i to v_j.*

In particular, if G is a *simple* graph, then the entry in the ith row and the jth column of the adjacency matrix will be 1 if v_i and v_j are adjacent, and 0 if they are not. In addition, the elements on the principal diagonal of the matrix (from the top left to the bottom right corner) will all be 0, because simple graphs do not have loops.

EXAMPLE 10.3.1

Construct the adjacency matrix for the graph shown in Figure 10.10, assuming the vertices are given in the order a, b, c, d, e.

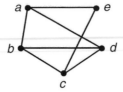

Figure 10.10

SOLUTION

$$\begin{array}{c c} & \begin{array}{c c c c c} a & b & c & d & e \end{array} \\ \begin{array}{c} a \\ b \\ c \\ d \\ e \end{array} & \left[\begin{array}{c c c c c} 0 & 1 & 0 & 1 & 1 \\ 1 & 0 & 1 & 1 & 0 \\ 0 & 1 & 0 & 1 & 1 \\ 1 & 1 & 1 & 0 & 0 \\ 1 & 0 & 1 & 0 & 0 \end{array}\right] \end{array}$$

EXAMPLE 10.3.2

Draw the graph with the following adjacency matrix:

$$\begin{array}{c c} & \begin{array}{c c c c} a & b & c & d \end{array} \\ \begin{array}{c} a \\ b \\ c \\ d \end{array} & \left[\begin{array}{c c c c} 0 & 0 & 0 & 1 \\ 0 & 0 & 2 & 0 \\ 0 & 2 & 0 & 1 \\ 1 & 0 & 1 & 1 \end{array}\right] \end{array}$$

SOLUTION

The graph is shown in Figure 10.11.

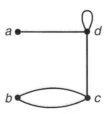

Figure 10.11

The entry in the *i*th row and the *j*th column of an adjacency matrix must be the same as the entry in the *j*th row and the *i*th column, because both entries represent the number of edges between v_i and v_j. We express this fact by saying that the adjacency matrix of any graph is *symmetric*. (Recall that we are dealing only with *undirected* graphs; the adjacency matrix of a directed graph need not be symmetric.) It follows that it is sufficient to give the entries on and below the principal diagonal of the matrix; there is no need to write down the entire matrix. This representation is called the *lower triangular* matrix representation of the graph. For example, the lower triangular matrix representation of the graph in Example 10.3.2 is:

$$
\begin{array}{c} \\ a \\ b \\ c \\ d \end{array}
\begin{array}{cccc} a & b & c & d \end{array}
\left[\begin{array}{cccc}
0 & & & \\
0 & 0 & & \\
0 & 2 & 0 & \\
1 & 0 & 1 & 1
\end{array} \right]
$$

Notice that the adjacency matrix of a graph depends on the order in which the vertices are labelled. A graph can have many adjacency matrices, corresponding to the different ways in which the vertices can be arranged. Using the theory that was introduced in Chapter 9, we know that there are $n!$ different ways in which n items can be arranged, so a graph with n vertices could have up to $n!$ different adjacency matrices.

10.4 *Isomorphism of graphs*

We want to look more closely now at the problem of how to tell whether graphs are isomorphic. We have said that two graphs are isomorphic if one of them can be 'moved around' until it looks like the other one (subject to certain rules, such as not breaking edges). This is a bit too vague to be really satisfactory, so we will now look for a more precise definition. In order to avoid some rather messy details that arise in the general case, we will assume in this section that we are dealing only with simple graphs.

The key to defining 'isomorphic' precisely is the idea of matching up the corresponding vertices in the two graphs. We did this in Figure 10.5, where we labelled corresponding vertices with the same letter. If two graphs are isomorphic, then we can associate with each vertex of one graph a corresponding vertex in the other graph.

Suppose we have two simple graphs G and H. Let $V(G)$ and $V(H)$ denote the vertex sets of G and H respectively. If G and H are isomorphic, then we should be able to show this by associating with each element of $V(G)$ the corresponding element of $V(H)$.

We have seen something like this before. Recall that in Chapter 6, we encountered the idea of associating each element of a set with an element of another set. In a word, what we have here is a *function*, with domain $V(G)$ and codomain $V(H)$:

$$ f: V(G) \rightarrow V(H) $$

There are some conditions that the function f must satisfy. We cannot associate two different vertices of G with the same vertex of H, nor can there be any vertices of H that are not associated with vertices of G. This means that f must be one-to-one and onto. There is a further requirement — if two vertices are adjacent in G, then the corresponding vertices must be adjacent in H, while if two vertices are not adjacent in G, then the corresponding vertices must not be adjacent in H. A function that satisfies all of these conditions is called an *isomorphism* from G to H.

Definitions

Let G and H be two simple graphs, with vertex sets $V(G)$ and $V(H)$ respectively. An *isomorphism* from G to H is a function $f: V(G) \rightarrow V(H)$ with the following properties:

- f is one-to-one and onto.
- For any two vertices u and v of G, if u and v are adjacent in G, then $f(u)$ and $f(v)$ are adjacent in H, and if u and v are not adjacent in G, then $f(u)$ and $f(v)$ are not adjacent in H.

If there exists an isomorphism from G to H, then G and H are *isomorphic*.

EXAMPLE 10.4.1

Find an isomorphism between the two graphs shown in Figure 10.12.

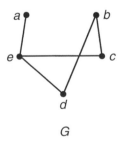

Figure 10.12

SOLUTION

One isomorphism is:

$$f: V(G) \rightarrow V(H), \quad f(a) = e, f(b) = b, f(c) = d, f(d) = a, f(e) = c$$

It is a straightforward exercise to check that this function satisfies the conditions of an isomorphism.

How can two graphs G and H be shown *not* to be isomorphic, if this is the case? It would be impractical to test all the functions from $V(G)$ to $V(H)$ to show that none of them are isomorphisms. A better way of showing that two graphs are not isomorphic is to find a graph-theoretic property that one graph has but the other does not. A 'graph-theoretic' property is a property that is retained if the graph is rearranged according to the rules we stated in Section 10.2.

For example, rearranging a graph in this way will not change the number of edges or the number of vertices. It follows that two graphs with different numbers of edges or vertices cannot possibly be isomorphic.

If two graphs have the same number of edges and the same number of vertices, then the problem is more difficult. In such cases, it is often useful to look at the degrees of the vertices. For example, if one graph has two vertices with degree 3, and the other has only one, then the graphs are not isomorphic. Sometimes even this test fails, and it is necessary to look for another property that distinguishes the two graphs.

EXAMPLE 10.4.2

Show that the pairs of graphs shown in Figures 10.13(a) and 10.13(b) are not isomorphic.

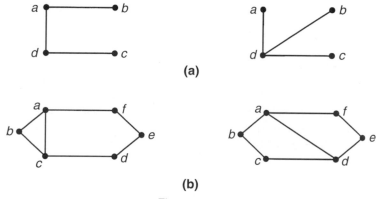

(a)

(b)

Figure 10.13

SOLUTION

(a) Both graphs in Figure 10.13(a) have four vertices and three edges. However, the second graph has a vertex with degree 3, while the first does not, so the graphs are not isomorphic. (Note that it would be incorrect to say that they are not isomorphic because vertex a has degree 2 in the first graph and degree 1 in the second. An isomorphism from one graph to another does not necessarily associate vertices that happen to be labelled with the same letter.)

(b) Both graphs in Figure 10.13(b) have six vertices and seven edges. Also, both graphs have four vertices with degree 2 and two vertices with degree 3. However, the first graph has a sequence *def* of three vertices with degree 2, with *d* adjacent to *e* and *e* adjacent to *f*, whereas there is no such sequence of three vertices with degree 2 in the second graph. Therefore the graphs are not isomorphic.

In the spirit of the earlier chapters, you might imagine that we would provide an algorithm for testing whether two graphs are isomorphic. Unfortunately, even the best known graph isomorphism algorithms are inefficient when applied to moderately large graphs. It is not known whether it is possible to construct a graph isomorphism algorithm that works 'efficiently' (in a sense that will be made precise in Chapter 13).

The idea that two mathematical objects (whether they are graphs or something else) can be fundamentally the same, even if they 'look different', is a very powerful one. We met a related idea in Chapter 8, where we saw that propositional logic and sets obey the 'same' laws of Boolean algebra. In this more general sense, isomorphism is one of the most important concepts in mathematics, because it allows us to ignore the superficial aspects of a problem and study its underlying mathematical structure.

10.5 *Paths and circuits*

If you look at a fairly detailed map of eastern Europe, you will see that there is a small territory on the coast of the Baltic Sea between Poland and Lithuania, called the Kaliningrad Region. The region belongs politically to the Russian Federation (although it is physically separate), and its main town is the Russian naval base of Kaliningrad. Until the Second World War, Kaliningrad was known by the name of Königsberg, and was in the part of Germany known as East Prussia.

The town of Königsberg provides the historical setting for the *Königsberg bridge problem*. This problem is one of the most famous in the history of mathematics, and its solution in about 1735 by the famous Swiss mathematician Leonhard Euler[1] (1707–1783) is generally regarded as marking the beginning of graph theory as a subject.

Two streams, called the Old Pregel and the New Pregel, join in the town to form the river Pregel (now called the Pregolya), but with the added complication that there is an island in the stream where they meet. In the early 18th century, the four areas of land thus formed (the north and south banks, the island, and the land between the two streams) were joined by seven bridges, as shown in Figure 10.14.

---

[1]Pronounced 'oiler'.

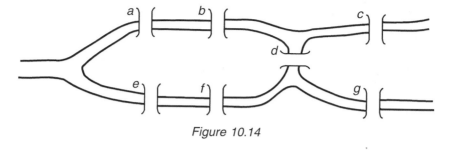

Figure 10.14

The problem that is now known as the Königsberg bridge problem must have first occurred to someone who was taking a stroll around the town. It goes as follows: is it possible for someone to take a walk around Königsberg in such a way as to cross each of the seven bridges *exactly once*?

A good way to get a feel for the problem is to try to find such a path. For example, if we start on the north bank, we could cross bridge *a* to the island, then *e* to the south bank. After returning to the island via *f*, we could then cross *d*, followed by *c* and *b*. But now we're in trouble; bridge *g* remains uncrossed, but we have no way of reaching it without crossing one of the bridges a second time.

At this point, you are encouraged to explore the problem yourself before reading further. We will obtain the answer later in this section. For now, we just make the observation that the Königsberg bridge problem can be treated as a problem in graph theory.

In order to see this, notice that in order to explore the problem, it is not necessary to know the precise geography of the pieces of land, the shape of the river, or the lengths of the bridges. In fact, we could collapse each of the four pieces of land down to a single point, without changing the problem in any essential way. All that matters is which pieces of land are joined by bridges, and by how many. If we represent the pieces of land by vertices and the bridges by edges joining those vertices, we obtain the graph shown in Figure 10.15.

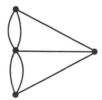

Figure 10.15

In the terminology of graph theory, the Königsberg bridge problem now asks whether it is possible to trace along the edges of the graph in Figure 10.15 in a continuous sequence (which we will be calling a *path*) in a way that ensures that each edge is traversed exactly once.

If the recreational needs of the citizens of 18th century Königsberg seem irrelevant to today's problems, imagine instead that the edges represent the streets on a mail delivery run. It is necessary to travel along every street in order to deliver the mail, but in the interests of efficiency we would prefer not to travel along any street a second time.

Before we can investigate the Königsberg bridge problem, we need to establish some more terminology.[2]

Definitions

A *path* of length n in a graph is a sequence of vertices and edges of the form:

$$v_0, e_1, v_1, e_2, v_2, ..., e_n, v_n$$

where e_i is an edge joining v_{i-1} and v_i, for all $i \in \{1, 2, ..., n\}$.
(In other words, e_1 is an edge joining v_0 and v_1, e_2 is an edge joining v_1 and v_2, and so on.)
A path for which $v_0 = v_n$ is called a *circuit*.

A path can be thought of as a sequence of vertices and edges that can be traced without lifting your pen off the paper. A path may include repeated edges or repeated vertices. The length of a path is the number of edges in the path. A circuit is a path that starts and ends at the same vertex.

A careful reading of the definition reveals that any vertex by itself is both a path and a circuit of length 0.

For example, consider the graph shown in Figure 10.16.

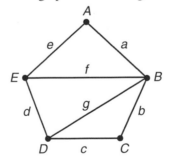

Figure 10.16

In this graph, A, a, B, g, D, d, E is a path of length 3, and $C, c, D, g, B, b, C, c, D$ is a path of length 4. $E, d, D, g, B, b, C, c, D, d, E$ is a circuit of length 5.

---

[2]Unfortunately, the terminology in graph theory is not fully standardised, and some terms, in particular 'path' and 'circuit' (defined here) and 'cycle' (defined in Chapter 11), are defined differently elsewhere. You should keep this in mind when referring to other textbooks.

When we write down a path, it is usually sufficient to list the edges; for example, the first of the paths listed above could be written *agd*.

We can now give a more precise definition of what it means for a graph to be connected.

Definition

A graph *G* is *connected* if, for any pair of vertices *u* and *v* of *G*, there is a path from *u* to *v*.

The graph shown in Figure 10.17 is not connected, because there is no path from *u* to *v*.

Figure 10.17

We can define a relation R on the set of vertices of a graph by the rule: $u R v$ if there is a path from u to v. Then R is an equivalence relation (you are asked to show this in Exercise 14). The equivalence classes for this relation are called the *components* of the graph; they are simply the connected 'pieces' of the graph. A connected graph is thus a graph that has only one component.

The path we are looking for in the Königsberg bridge problem has an additional property — it must include every edge of the graph exactly once. More generally, we can ask the following question: Given *any* connected graph, can we find a path that includes every edge exactly once? (If a graph is not connected, there could still be a path that includes every edge exactly once, but only in the not particularly interesting case where all the components except one consist of just one vertex. For this reason, we will restrict our attention to connected graphs.)

Definitions

Let *G* be a connected graph. An *Eulerian path* is a path that includes every edge of *G* exactly once. An *Eulerian circuit* is an Eulerian path for which the first and the last vertices co-incide (i.e. an Eulerian circuit is an Eulerian path that is also a circuit).

A connected graph that has an Eulerian circuit is called *Eulerian*. A connected graph that has an Eulerian path but no Eulerian circuit is called *semi-Eulerian*.

Note that an Eulerian circuit may have repeated vertices.

Figure 10.18 shows the graph for the Königsberg bridge problem again, this time with the vertices labelled.

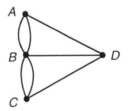

Figure 10.18

The problem can now be restated as follows: Does this graph have an Eulerian path (i.e. is it either an Eulerian or a semi-Eulerian graph)?

Suppose we focus our attention on vertex *A* for a moment. If there is an Eulerian path in the graph, then at some stage it must enter *A* along one of the three edges incident to *A*, and leave via another one. That leaves just one unused edge incident to *A*. It would be impossible to pass through *A* a second time without repeating an edge that has already been used, so the remaining edge can be used only if the path either starts or ends at *A*. This rules out any possibility of an Eulerian circuit, but there could still be an Eulerian path.

Now consider vertex *C*. Like vertex *A*, this vertex also has degree 3, so the reasoning that we have just applied to *A* applies to *C* too. We deduce that any Eulerian path would have to start or end at *C*.

So far, it appears that there could be an Eulerian path, provided that it starts at *A* and ends at *C* (or vice versa). But now, if we apply the same reasoning to vertex *D* (which also has degree 3), we see that any Eulerian path would have to start or end at *D*. We conclude that no Eulerian path exists, and therefore that it is impossible to find a route around Königsberg that crosses every bridge exactly once.

The solution to the Königsberg bridge problem suggests a more general result about Eulerian paths and circuits in connected graphs. The key to the reasoning we used was to look at the degrees of the vertices. If a vertex has *even* degree, say $2n$, then an Eulerian path will use all the edges incident to the vertex by entering and leaving the vertex n times. If a vertex has *odd* degree, on the other hand, then there will be one edge left over which can be used only if the path starts or ends there. Therefore, any connected graph with more than two vertices with odd degree cannot have an Eulerian path. This conclusion forms part of the following theorem.

Theorem

Let G be a connected graph.

1. If all the vertices of G have even degree, then G is Eulerian.
2. If exactly two vertices of G have odd degree, then G is semi-Eulerian, and every Eulerian path in G must start at one of the vertices with odd degree and end at the other.
3. If G has more than two vertices with odd degree, then G is neither Eulerian nor semi-Eulerian.

EXAMPLE 10.5.1

Classify each of the graphs in Figure 10.19 as Eulerian, semi-Eulerian, or neither, and find an Eulerian path or an Eulerian circuit if one exists.

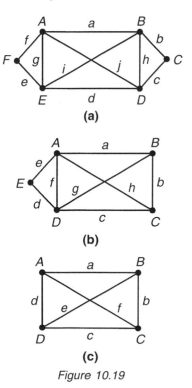

(a)

(b)

(c)

Figure 10.19

Solution

(a) All the vertices have even degree, so the graph is Eulerian. An Eulerian circuit is *abcdefgihj*.

(b) Two vertices, *B* and *C*, have odd degree, so it is possible to find an Eulerian path from *B* to *C*. One such path is *aedcbgfh*.

(c) There are four vertices with odd degree, so this graph does not have an Eulerian path.

We have not yet proved Parts 1 and 2 of the theorem. We will establish Part 1 by devising an algorithm for finding an Eulerian circuit in a connected graph in which all the vertices have even degree. The fact that the algorithm always works amounts to a proof of Part 1. Part 2 will follow as a corollary.

Consider again the graph in Part (a) of Example 10.5.1. Suppose we decide to begin our Eulerian circuit at vertex *F*. (Because an Eulerian circuit is a circuit, we should be able to start at any vertex.) If we follow the circuit *fge*, we then find that we cannot go any further without repeating edges, yet we have not used all the edges in the graph.

The key to the algorithm for finding an Eulerian circuit is that we can extend the circuit we already have, by inserting extra edges and vertices at any vertex where there are unused edges. *A* is such a vertex. Starting at *A*, and following unused edges until we can go no further, we produce the circuit *ahj*. This new circuit can be inserted into the original one to produce the extended circuit *fahjge*.

The circuit we have obtained so far still doesn't include all of the edges of the graph, so we repeat the extension process. There is a vertex, *B*, on the circuit, where there are still some unused edges. Starting at *B*, and using only edges that have not previously been used, we obtain the circuit *bcdi*. Inserting the new circuit into the existing one, we obtain *fabcdihjge*. This circuit uses all of the edges, so it is an Eulerian circuit.

Why does this procedure work? To answer this question, we need to explain why, when a path is being created for insertion into a circuit that has already been obtained, we can be sure that the new path will always be a circuit.

The explanation runs as follows. Since every vertex of the graph has even degree, the new path can always leave any vertex it has entered, with the exception only of the vertex where it started. This guarantees that each new path will always be a circuit, and so can be inserted into the circuit previously obtained.

We will now write this procedure as an algorithm in a form suited to hand computation. (In a computer-oriented version of the algorithm, the graph would be input as an adjacency matrix, and some of the steps would need to be specified in greater detail.)

1. Input an Eulerian graph G, with vertex set $V(G) = \{v_1, v_2, \ldots, v_m\}$ and edge set $E(G) = \{e_1, e_2, \ldots, e_n\}$.[3]
2. *circuit* $\leftarrow v_1$; *unused_edges* $\leftarrow E(G)$

{*circuit* is a circuit (expressed as a sequence of vertices and edges) that is built up piece by piece until it forms an Eulerian circuit. Initially it consists of a single vertex. The set *unused_edges* keeps track of the edges that have not yet been used.}

{Step 3 is the main part of the algorithm. Each time the **While-do** loop is executed, a new circuit is formed and inserted into *circuit* in place of the vertex *insertion_point*.}

3. **While** *unused_edges* $\neq \varnothing$ **do**
    3.1. *insertion_point* $\leftarrow$ first vertex in circuit with unused edges incident to it
    3.2. $v \leftarrow$ *insertion_point*; *new_circuit* $\leftarrow v$

{Each time the **Repeat-until** loop in Step 3.3 is executed, an edge and a vertex are appended to *new_circuit*.}

    3.3. **Repeat**
        3.3.1. $e \leftarrow$ first element of *unused_edges* incident to v
        3.3.2. $v \leftarrow$ vertex adjacent to v via edge e
        3.3.3. *new_circuit* $\leftarrow$ *new_circuit*, e, v
        3.3.4. *unused_edges* $\leftarrow$ *unused_edges* $- \{e\}$
        until no element of *unused_edges* is incident to v
    3.4. *circuit* $\leftarrow$ (*circuit* before *insertion_point*), *new_circuit*, (*circuit* after *insertion_point*)
4. Output *circuit*

EXAMPLE 10.5.2

Trace the algorithm using the graph shown in Figure 10.20 as input, with vertex set $\{A, B, C, D\}$ and edge set $\{p, q, r, s, t, u\}$.

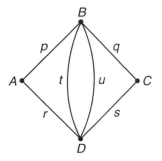

Figure 10.20

---

[3]There is a notational ambiguity here; we have been using braces in algorithms to denote comments, but braces are also used to write sets in enumerated or predicate form. The problem is not serious, as the meaning will always be clear from the context.

SOLUTION

A full trace would be rather tedious here, so we will take a short-cut and treat the four steps within Step 3.3 as a single step (see Table 10.1).

Step	circuit	insertion _point	e	v	new_circuit	unused _edges	Output
2	A	–	–	–	–	$\{p, q, r, s, t, u\}$	–
3.1	A	A	–	–	–	$\{p, q, r, s, t, u\}$	–
3.2	A	A	–	A	A	$\{p, q, r, s, t, u\}$	–
3.3.1–4	A	A	p	B	ApB	$\{q, r, s, t, u\}$	–
3.3.1–4	A	A	q	C	ApBqC	$\{r, s, t, u\}$	–
3.3.1–4	A	A	s	D	ApBqCsD	$\{r, t, u\}$	–
3.3.1–4	A	A	r	A	ApBqCsDrA	$\{t, u\}$	–
3.4	ApBqCsDrA	A	r	A	ApBqCsDrA	$\{t, u\}$	–
3.1	ApBqCsDrA	B	r	A	ApBqCsDrA	$\{t, u\}$	–
3.2	ApBqCsDrA	B	r	B	B	$\{t, u\}$	–
3.3.1–4	ApBqCsDrA	B	t	D	BtD	$\{u\}$	–
3.3.1–4	ApBqCsDrA	B	u	B	BtDuB	∅	–
3.4	ApBtDuBqCsDrA	B	u	B	BtDuB	∅	–
4	ApBtDuBqCsDrA	B	u	B	BtDuB	∅	ApBtDuBqCsDrA

| Table 10.1

If a graph is semi-Eulerian, we can find an Eulerian path in the following way:

1. Insert a new edge between the two vertices with odd degree. (The resulting graph will be Eulerian, because all of the vertices will have even degree.)
2. Let the vertex set and the edge set of the Eulerian graph be $\{v_1, v_2, \ldots, v_m\}$ and $\{e_1, e_2, \ldots, e_n\}$ respectively, where the new edge e_1 is incident to v_1 and v_2. Apply the algorithm to obtain an Eulerian circuit.
3. Remove v_1, e_1 from the beginning of the Eulerian circuit. What remains is an Eulerian path joining the two vertices with odd degree.

Now that we have obtained a satisfactory solution to the problem of determining whether a graph has a path or a circuit that includes every edge exactly once, we turn to what appears to be a similar problem — to determine whether a graph has a path or a circuit that includes every *vertex* exactly once.

Definitions

Let *G* be a connected graph. A *Hamiltonian[4] path* is a path that includes every vertex of *G* exactly once, except only that the first and the last vertices may coincide. A *Hamiltonian circuit* is a Hamiltonian path for which the first and the last vertices coincide (i.e. a Hamiltonian circuit is a Hamiltonian path that is also a circuit).

A connected graph that has a Hamiltonian circuit is called *Hamiltonian.*

Note that a Hamiltonian path or a Hamiltonian circuit need not use all of the edges of the graph.

For example, consider the two graphs shown in Figure 10.21.

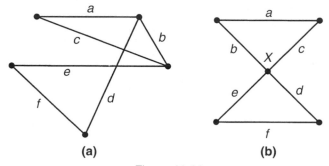

Figure 10.21

The first graph (Figure 10.21(a)) is Hamiltonian, because it has a Hamiltonian circuit, *cefda*. The second graph ((b)) has a Hamiltonian path, *abef*; however, the graph is not Hamiltonian, because any circuit passing through all of the vertices would have to pass through vertex *X* twice.

The problem of determining whether a graph is Hamiltonian appears at first sight to be similar to the problem of determining whether a graph is Eulerian, and it might be imagined that it could be solved in a similar way. In fact, no efficient algorithm for determining whether a graph is Hamiltonian has so far been produced, and it may well turn out to be the case that no such algorithm can possibly exist. This problem is one of a certain class of problems, all of which can be shown to be equivalent, in the sense that if an efficient algorithm could be found for solving just one of the problems, then there would be an efficient algorithm for solving all of them. The problem of the 'Travelling sales representative', which we will examine in Chapter 11, is another such problem.

---

[4]Hamiltonian paths and circuits are named after the Irish mathematician Sir William Hamilton (1805–1865). In 1859, Hamilton produced a puzzle consisting of a dodecahedron (a regular solid with 12 faces) on which each vertex was labelled with the name of a city. The problem was to find a way of travelling along the edges of the dodecahedron, visiting each city exactly once.

Exercises

1. Explain how each of the following situations could be modelled as a graph. In each case, state what meaning (if any) can be attached to loops and parallel edges, and whether it would make sense to use directed edges.

   (a) An electronic circuit.
   (b) A chemical molecule.
   (c) A group of people and the relationship of friendship between two people.
   (d) The management structure of a company.
   (e) The modules of a computer program, and the relationship of one module calling another.

2. Draw all 11 simple graphs with four vertices.

3. Verify for each of the graphs shown in Figure 10.22 that the sum of the degrees of the vertices equals twice the number of edges.

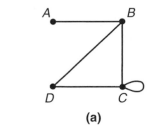

(a)

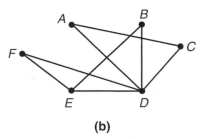

(b)

Figure 10.22

4. Draw a graph whose vertices have the following degrees, or explain why no such graph exists:

   (a) 2, 3, 3, 4, 5   (b) 2, 3, 3, 3, 3

5. Obtain an alternative proof that a complete graph with n vertices has $\dfrac{n(n-1)}{2}$ edges, using the result that the sum of the degrees of the vertices equals twice the number of edges.

6. If a simple graph G has n vertices and m edges, how many edges does $\overline{G}$ have?

7. Construct an adjacency matrix for each of the graphs shown in Figure 10.23.

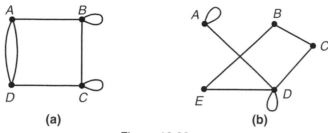

(a)   **(b)**

Figure 10.23

8. Draw the graphs with the following adjacency matrices:

(a) $\begin{bmatrix} 0 & 1 & 1 & 0 \\ 1 & 0 & 0 & 1 \\ 1 & 0 & 0 & 1 \\ 0 & 1 & 1 & 0 \end{bmatrix}$ (b) $\begin{bmatrix} 0 & 0 & 1 & 2 & 1 \\ 0 & 1 & 1 & 1 & 0 \\ 1 & 1 & 0 & 0 & 1 \\ 2 & 1 & 0 & 0 & 1 \\ 1 & 0 & 1 & 1 & 2 \end{bmatrix}$

9. Write down the lower triangular matrix representation of each of the graphs in Exercise 7.

10. How can the degrees of the vertices of a graph be determined directly from the adjacency matrix, without drawing the graph? (Answer the question first for the simpler case where the graph has no loops.)

11. Show that the pairs of graphs in Figure 10.24 are isomorphic, by finding an isomorphism from one to the other.

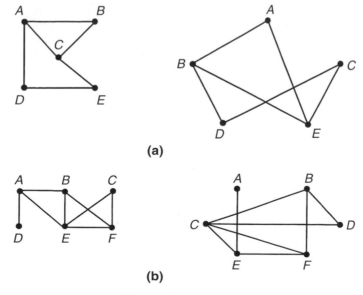

(a)

(b)

Figure 10.24

12. Show that the pairs of graphs in Figure 10.25 are not isomorphic.

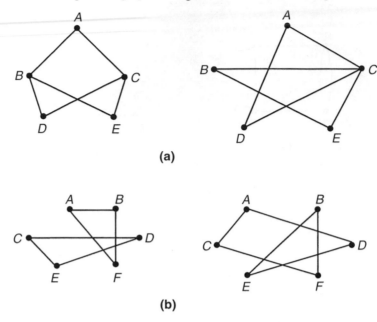

(a)

(b)

Figure 10.25

13. (a) Draw the complement $\overline{G}$ of graph G (Figure 10.26), and show that G and $\overline{G}$ are isomorphic.

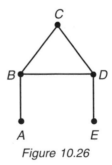

Figure 10.26

(b) Could a simple graph with six vertices be isomorphic to its complement? (Use the result of Exercise 6.)

14. Let G be a graph, and let R be the relation on the set of vertices of G defined by the rule: $u\,R\,v$ if there is a path from u to v. Prove that R is an equivalence relation.

15. Classify the graphs in Figure 10.27 as Eulerian, semi-Eulerian, or neither, and find an Eulerian path or an Eulerian circuit if one exists.

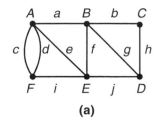

(a)

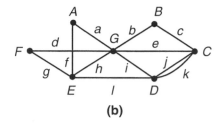

(b)

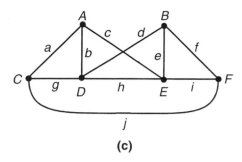

(c)

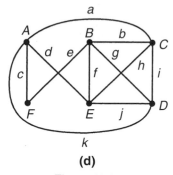

(d)

Figure 10.27

16. The theorem for determining whether a graph is Eulerian, semi-Eulerian or neither makes no mention of the case of a graph with exactly one vertex with odd degree. Why not?

17. For each of the graphs in Exercise 15, find a Hamiltonian circuit or explain why no Hamiltonian circuit exists.

18. Prove that if G is any simple graph, then at least one of G and $\overline{G}$ is a connected graph. (Assume G is not connected, and prove that there is a path between any two vertices u and v of $\overline{G}$ by considering two cases: u and v are in the same component of G, and u and v are in different components.)

19. Prove that if G is a simple graph with at least two vertices, then it is always possible to find two vertices of G with the same degree. (Begin by listing all the possible values that the degree of a vertex can take in a simple graph with n vertices.)

11. TREES

11.1 Introduction to trees

In this chapter we will study the particular type of graph known as a tree, and investigate some applications of trees to practical problems. Trees arise in the problem of designing a communications network as cheaply as possible, and in the representation of hierarchical structures and decision processes, as well as in many other situations.

As we did in Chapter 10, we will assume that all of the graphs we are dealing with are finite.

We begin with a definition.

Definition

A *cycle* is a path in a graph with the following properties:

- It includes at least one edge.
- There are no repeated edges.
- The first and last vertices coincide, but there are no other repeated vertices.

A cycle with *n* edges is called an *n-cycle*.

A cycle is thus a circuit with some additional properties. A vertex by itself is not a cycle, because a cycle must include at least one edge. If two cycles consist of the same vertices and edges, we will regard them as the same cycle; for example, a cycle with edges $e_1e_2e_3e_4$ is the same cycle as $e_2e_3e_4e_1$ and $e_3e_2e_1e_4$.

A 1-cycle is a loop, while a 2-cycle is formed by a pair of parallel edges. The shortest possible cycle in a simple graph is a 3-cycle.

EXAMPLE 11.1.1

List the cycles in the graph shown in Figure 11.1.

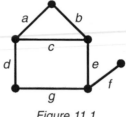

Figure 11.1

SOLUTION

There are three cycles: *abc*, *cegd* and *abegd*.

Definition

A *tree* is a connected graph with no cycles.

Some examples of trees are shown in Figure 11.2.

Figure 11.2

If the word 'connected' is removed from the definition of a tree, we are left with a definition for a graph that has no cycles but is not necessarily connected. Such a graph is called a *forest*; a forest consists of one or more components, each of which is a tree.

Three obvious properties of trees are listed below:

- Inserting an edge between any two vertices of a tree produces a graph containing a cycle.
- Removing any edge from a tree produces a disconnected graph.
- For each pair of vertices in a tree, there is exactly one path from one vertex to the other that does not repeat any vertices or edges.

The property given in the following theorem is less obvious.

Theorem

Any tree with n vertices has $n-1$ edges.

It is easily checked that the theorem is true for each of the trees shown in Figure 10.2. For example, the rightmost tree has 13 vertices and 12 edges.

We will prove the theorem using induction on n. The proof is an interesting example of induction because it shows that the method can be applied more widely than just to the algebraic problems we dealt with in Chapter 7.

In the course of the proof, we will need to use the following fact, which should be obvious: in any tree with at least two vertices, it is always possible to find a vertex with degree 1.

For the base step of the proof, we simply note that a tree with one vertex must have no edges, because a tree cannot contain loops.

We begin the inductive step by assuming that any tree with k vertices has $k-1$ edges. Let T be a tree with $k+1$ vertices. 'Prune' the tree T by removing a vertex v with degree 1 and the edge e incident to v. The resulting graph is a tree with k vertices, and by the inductive hypothesis it has $k-1$ edges. Now restore T by replacing v and e. In the process, the number of edges increases by 1, so T has k edges. This completes the inductive step.

It follows by induction that a tree with n vertices has $n-1$ edges, for all values of n. This concludes the proof.

We will use this result in the minimal spanning tree algorithm in the next section.

11.2 *Local area networks and minimal spanning trees*

In this section, we will solve a practical problem in the design of local area networks.

Suppose there is a site, such as a university campus, on which there are several buildings containing computer laboratories. We would like to build a communications network, called a *local area network* (or LAN), in such a way that it is possible for a computer in any laboratory to communicate with a computer in any other laboratory. Of course, this does not mean that every pair of buildings must be directly linked; it is quite acceptable for two buildings to be linked indirectly via a third one, for example. Given that installing a link between two buildings is a costly process, we would like to build the network as cheaply as possible.

For example, suppose we have four buildings: A, B, C and D. The costs of linking each pair of buildings (in thousands of dollars) are tabulated in Table 11.1.

	A	B	C	D
A	–			
B	120	–		
C	100	105	–	
D	140	125	90	–

Table 11.1

We can represent the situation by means of a graph in which the vertices are the buildings (Figure 11.3).

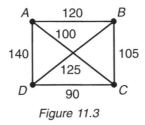

Figure 11.3

A graph of this type is called a *weighted* graph.

Definitions

A *weighted graph* is a simple graph in which each edge has a positive number attached to it. The number attached to an edge in a weighted graph is called the *weight* of the edge.

In the present example, the weights are the costs of installing the links. In other problems in which weighted graphs are used, the weights might represent road distances, signal travelling times, or link capacities.

We do not need to install all of the links in our example; it is sufficient to install just the three cheapest ones: *CD*, *BC* and *AC*. With these links in place, any two laboratories can communicate with one another.

This example was small enough that the solution could be obtained easily by inspection. In a problem involving a large number of sites, it is not practical to determine the solution by trial and error in this way. For this reason, we now turn our attention to developing an algorithm for finding the cheapest network linking the sites.

As a first step towards solving the problem, we can look for some properties that the solution must have. The solution is itself a graph, consisting of all of the vertices but only some of the edges of the original weighted graph. Clearly the solution must be a connected graph if we want to ensure that any two sites can communicate with each other. Furthermore, since the solution represents the cheapest possible network connecting the sites, it cannot contain any cycles because if a cycle were present, then one of the edges in the cycle could be removed to produce a connected graph with a smaller total weight.

We can sum up these observations as follows. The solution to the problem must be a tree. The vertices of the tree must be the vertices of the original graph, and the edges of the tree must be selected from the set of edges in the original graph. A tree that satisfies these requirements is called a *spanning tree* of the graph.

Typically, a connected graph will have many spanning trees. In Figure 11.3, not only do the edges *CD*, *BC* and *AC* form a spanning tree, but so do the edges *AB*, *AC* and *AD*, and the edges *AB*, *BC* and *CD*. The problem is to find a spanning tree with the smallest possible total weight, since this will represent the cheapest possible network. A spanning tree with this property is called a *minimal spanning tree*.

Here, then, is the problem stated in the terminology of graph theory.

Minimal spanning tree problem

Design an algorithm that inputs a connected weighted graph and outputs a minimal spanning tree for the graph.

How might we go about finding a minimal spanning tree? A plausible approach would be to start at one of the vertices, and pick an edge with the least weight of all the edges incident to that vertex. This edge forms the start of our minimal spanning tree. Next, we could look for an edge with least weight of all the edges that join up with the edge we already have. Continuing in this way, adding one edge at a time, we might hope to obtain a minimal spanning tree.

We would have to make one modification to this procedure. If the addition of an edge would create a cycle, then clearly we would not want to add that edge. What we need to do at each step, then, is to *add an edge of least weight that does not form a cycle*.

The procedure we have just described can be formulated as an algorithm, known as *Prim's algorithm*:

1. Input a connected weighted graph *G*, with vertex set
   $V(G) = \{v_1, \ldots, v_n\}$ and edge set $E(G) = \{e_1, \ldots, e_m\}$
2. $T \leftarrow \{v_1\}$; *unused_edges* $\leftarrow E(G)$
3. **For** $i = 1$ **to** $n - 1$ **do**
   3.1. $e \leftarrow$ the first edge with minimal weight in *unused_edges* that is incident to exactly one vertex in *T*
   3.2. $v \leftarrow$ the vertex not in *T* to which *e* is incident
   3.3. $T \leftarrow T \cup \{e, v\}$
   3.4.   *unused_edges* $\leftarrow$ *unused_edges* $- \{e\}$
4. Output *T*

Notice that the **For-do** loop is executed $n - 1$ times (where n is the number of vertices in the weighted graph), because any tree with n vertices has $n - 1$ edges, and a spanning tree for a graph has the same number of vertices as the original graph.

EXAMPLE 11.2.1

Use Prim's algorithm to find a minimal spanning tree for the weighted graph shown in Figure 11.4.

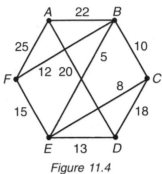

Figure 11.4

SOLUTION

Starting at vertex A, we find that the edge with least weight incident to A is the edge AD with weight 20, so this is the first edge in the minimal spanning tree T.

We look next at the edges incident to both A and D and not yet in T. Of these, DE has the least weight, so we add it to T.

In a similar fashion, we add EB and EC to T. At this stage, the edge with least weight incident to a vertex of T is BC with weight 10, but this edge is incident to two vertices of T, and therefore we cannot add it to T. The edge with least weight that is incident to exactly one vertex of T is BF, so this is the edge to be added.

The tree T now has a total of 5 edges. Since the original graph has 6 vertices, and $5 = 6 - 1$, there are no more edges to be added.

The minimal spanning tree is shown in Figure 11.5; it has a total weight of 58.

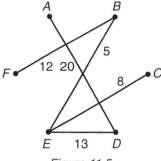

Figure 11.5

A trace table for the algorithm is shown in Table 11.2.

Step	e	v	T	unused_edges	Output
2	–	–	{A}	{AB, AD, AF, BC, BE, BF, CD, CE, DE, EF}	–
3.1–3.4	AD	D	{A, AD, D}	{AB, AF, BC, BE, BF, CD, CE, DE, EF}	–
3.1–3.4	DE	E	{A, AD, D, DE, E}	{AB, AF, BC, BE, BF, CD, CE, EF}	–
3.1–3.4	BE	B	{A, AD, D, DE, E, BE, B}	{AB, AF, BC, BF, CD, CE, EF}	–
3.1–3.4	CE	C	{A, AD, D, DE, E, BE, B, CE, C}	{AB, AF, BC, BF, CD, EF}	–
3.1–3.4	BF	F	{A, AD, D, DE, E, BE, B, CE, C, BF, F}	{AB, AF, BC, CD, EF}	–
4	BF	F	{A, AD, D, DE, E, BE, B, CE, C, BF, F}	{AB, AF, BC, CD, EF}	{A, AD, D, DE, E, BE, B, CE, C, BF, F}

| Table 11.2

While Prim's algorithm appears to be a reasonable way of finding a spanning tree with small total weight, it is not obvious that it always produces a minimal spanning tree. In fact, Prim's algorithm does always yield a minimal spanning tree in any connected weighted graph; the proof of this fact is not given here, but can be found in many textbooks on graph theory.

Prim's algorithm is one of a class of algorithms known as *greedy* algorithms. A greedy algorithm is an algorithm in which at each step we perform whatever action appears to be best at that step, without considering whether it leads to the best result overall. In each iteration of the **For-do** in Prim's algorithm, we choose the lowest weight edge to add to the tree (subject to the restriction that no cycles are created). It turns out that this process leads to the lowest total weight in the spanning tree.

An example of a problem in which a greedy strategy doesn't always work is the Travelling sales representative (or Travelling salesman) problem. A sales representative wants to visit a number of towns and return home, travelling the shortest possible total distance in the process. If we represent the towns and roads by a weighted graph, in which the weights are the road distances, then the problem becomes: find the circuit with the smallest possible total weight passing through all of the vertices. (We could call this a minimal spanning *circuit*, as distinct from the minimal spanning tree of the previous problem.)

A greedy approach to the Travelling sales representative problem might work in the following way. Starting from the home vertex, travel at each step by the shortest available path to the nearest vertex that has not been visited already. When all of the vertices have been visited, return to the starting vertex by the shortest path. This method is known as the 'Nearest neighbour algorithm'.

If the Nearest neighbour algorithm is applied to the graph in Example 11.2.1 (with the weights now interpreted as distances, and with A as the home vertex), the result is the circuit *ADEBCBFA*, with a total distance of 95. This is not the best solution, however; for example, the circuit *ADCEBFA* is shorter, with a total distance of 88. Greed doesn't always pay!

We mention in passing that it is not known whether it is possible to design an efficient algorithm for the Travelling sales representative problem.

There is one further question that we would like to answer before we conclude this section. How can a weighted graph be represented in a form suitable for machine computation? The answer that most readily comes to mind is to use a matrix similar to an adjacency matrix, but with the entry in row i and column j containing the weight of the edge from vertex v_i to vertex v_j rather than the number of edges. (Recall that, by definition, every weighted graph is a simple graph, so there will be no multiple edges.) A matrix of this type is called a *weight matrix*.

What should we put in row i and column j if there is no edge from v_i to v_j? In order to answer this question, we need to realise that weighted graphs can be classified into two types. In the first type, the weights represent a penalty of some kind, such as a distance or a cost, which we would want to try to *minimise* in a typical application. The graphs in the Minimal spanning tree problem and the Travelling sales representative problem are of this type. In the second type, the weights represent something we would want to *maximise*, such as the capacity of a communications channel. In a graph of the first type, having two vertices with no connecting edge is even 'worse' than having them joined by an edge with a very large weight. This is indicated in the weight matrix by treating two non-adjacent vertices as if they were joined by an edge of infinite weight. For weighted graphs of the second type, a weight of zero is appropriate for non-adjacent vertices. The entries on the diagonal of the weight matrix are zero for both types of weighted graph. We will consider only the first type here.

Definition

Let G be a weighted graph with vertices v_1, v_2, ..., v_n, in which the weights represent a penalty of some kind. The *weight matrix* of G is the $n \times n$ matrix for which the entry w_{ij} in the ith row and the jth column is given by the rule:

$$w_{ij} = \begin{cases} 0 \text{ if } i = j \\ \text{the weight of the edge from } i \text{ to } j \text{ if } v_i \text{ and } v_j \\ \quad \text{are adjacent} \\ \infty \text{ if } i \neq j \text{ and } v_i \text{ and } v_j \text{ are not adjacent} \end{cases}$$

EXAMPLE 11.2.2

Write down the weight matrix for the weighted graph in Example 11.2.1.

SOLUTION

$$\begin{array}{c c} & \begin{array}{cccccc} A & B & C & D & E & F \end{array} \\ \begin{array}{c} A \\ B \\ C \\ D \\ E \\ F \end{array} & \left[\begin{array}{cccccc} 0 & 22 & \infty & 20 & \infty & 25 \\ 22 & 0 & 10 & \infty & 5 & 12 \\ \infty & 10 & 0 & 18 & 8 & \infty \\ 20 & \infty & 18 & 0 & 13 & \infty \\ \infty & 5 & 8 & 13 & 0 & 15 \\ 25 & 12 & \infty & \infty & 15 & 0 \end{array} \right] \end{array}$$

11.3 Minimal distance paths

Suppose we have a weighted graph that represents a communications network as in the previous section, but now the weight of each edge represents the time taken by a signal to travel along that link in the network. Given any two nodes (vertices) in the network, it is a problem of considerable practical importance to find the quickest path for a signal to take through the network from one node to the other.

In order to cast the problem in graph-theoretic terms, the following definitions are useful.

Definitions

Let G be a weighted graph, and let u and v be vertices in G. If P is a path from u to v, then the *length* of P is the sum of the weights of the edges in P. The *distance* from u to v is the smallest of the lengths of all the paths from u to v.

We will use the notation $d(u, v)$ to denote the distance from u to v. It will also be convenient to use the notation $weight(e)$ to denote the weight of an edge e.

Notice that the definition of the length of a path in a weighted graph is different from the definition in a graph without weights, which we encountered in Chapter 10. (Recall that in the latter case the length of a path is simply the number of edges in the path.)

In the application we have in mind, the weights actually represent times rather than distances. However, we can imagine that a weighted graph could also represent a road network in which the weights are the road distances.

The problem can now be stated in the terminology of graph theory as follows.

Minimum distance problem

Design an algorithm to input a weighted graph and two vertices u and v, and output the distance $d(u, v)$ from u to v together with the path from u to v that achieves this distance.

It turns out that it is no more difficult to design an algorithm to answer a slightly more general problem: given a weighted graph G and a vertex u, find the distances from u to all vertices v of G. Roughly speaking, the reason for this is that in the process of finding the distance from u to v, we also find the distance from u to each vertex along the way.

Dijkstra's algorithm is an algorithm that provides an efficient solution to the minimum distance problem. Just as we did with Prim's algorithm for finding a minimal spanning tree, we will present Dijkstra's algorithm in a form suited to hand computation.

1. Input a connected weighted graph G with vertex set
   $V(G) = \{v_1, ..., v_n\}$ and edge set $E(G) = \{e_1, ..., e_m\}$

   {We will obtain, in the form of a spanning tree, T, the minimum distance paths from v_1 to each of the other vertices. In the process, we will also obtain the distance of each vertex from v_1 as we add the vertex to T.}

2. $T \leftarrow \{v_1\}; d(v_1, v_1) \leftarrow 0$
3. **For** $j = 1$ **to** $n - 1$ **do**
   3.1. $w \leftarrow$ the first vertex for which $d(v_1, w) + weight(e)$ is minimised, where w ranges over all the vertices in T, and e ranges over all the edges in $G - T$ that are incident to w
   3.2. $e \leftarrow$ the first edge incident to w for which $d(v_1, w) + weight(e)$ is minimised
   3.3. $v \leftarrow$ the vertex in $G - T$ to which e is incident
   3.4. $T \leftarrow T \cup \{e, v\}$
   3.5. $d(v_1, v) \leftarrow d(v_1, w) + weight(w, v)$
4. Output T

EXAMPLE 11.3.1

Apply Dijkstra's algorithm to find the minimum distance from A to C in the weighted graph in Example 11.2.1, and the path that achieves this distance.

SOLUTION

Table 11.3 is a partial trace table for the algorithm.

Step	w	e	v	T	$d(v_1, v)$	Output
2	–	–	–	$\{A\}$	–	–
3.1–3.5	A	AD	D	$\{A, AD, D\}$	20	–
3.1–3.5	A	AB	B	$\{A, AD, D, AB, B\}$	22	–
3.1–3.5	A	AF	F	$\{A, AD, D, AB, B, AF, F\}$	25	–
3.1–3.5	B	BE	E	$\{A, AD, D, AB, B,$ $AF, F, BE, E\}$	27	–
3.1–3.5	B	BC	C	$\{A, AD, D, AB, B,$ $AF, F, BE, E, BC, C\}$	32	–
4	B	BC	C	$\{A, AD, D, AB, B,$ $AF, F, BE, E, BC, C\}$	32	$\{A, AD, D, AB, B,$ $AF, F, BE, E, BC, C\}$

Table 11.3

The distance from A to C is 32, and the path is ABC.

11.4 Rooted trees

In this section, we consider the kind of tree known as a *rooted tree*, and investigate one of its applications to a problem in computing.

A rooted tree is a tree in which one of the vertices is specified as the *root*. By convention, rooted trees are usually drawn with the root at the top. The vertices adjacent to the root (the 'first generation') are shown in a horizontal line below the root, the vertices that can be reached from the root by a path of length 2 (the 'second generation') are shown in a line below the first generation, and so on.

An example of a rooted tree is shown in Figure 11.6.

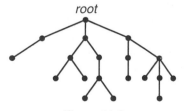

Figure 11.6

A family tree, showing the descendants of an individual, is a familiar everyday example of a rooted tree. For this reason, much of the terminology associated with family trees is carried over to rooted trees. For example, we refer to the vertex immediately above a given vertex as the *parent* of the given vertex, while a vertex immediately below a given vertex is a *child* of that vertex. Note that every vertex in a rooted tree except the root has exactly one parent. Similarly, we can say that one vertex in a rooted tree is an *ancestor* or a *descendant* of another, with the obvious meaning in each case. Changing the analogy, a vertex with no children is called a *leaf*.

The tree diagrams we encountered in Chapter 9 are rooted trees, although we drew them with the root on the left rather than at the top. In that chapter, tree diagrams were used to represent decision processes, in which we were faced with a choice at each of several steps in a procedure. This is one of the most widely used applications of rooted trees in computing.

As an example of a decision tree, suppose we want to sort three distinct numbers, denoted by a, b and c, into increasing order. In order to carry out the sorting process, we need to take the numbers two at a time and compare them. In some cases, two such comparisons will be sufficient to determine the order in which the three numbers should be arranged, while in other cases three comparisons will be needed. The procedure is depicted in the rooted tree shown in Figure 11.7. The leaves represent the final arrangements of the three numbers into increasing order, and the other vertices represent points at which a decision is made. The decision process begins at the root, and moves down the tree until a leaf is reached.

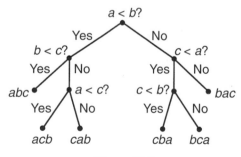

Figure 11.7

The rooted tree in this example has the property that every vertex that is not a leaf has exactly two children. A rooted tree with this property is called a *binary rooted tree*. In a binary rooted tree, the two children of each parent are identified as the *left child* and the *right child*, according to where they are placed when the tree is drawn. For any given parent vertex, the left child is the root of the *left subtree*, the binary rooted tree consisting of the left child and its descendants. Similarly, the right child is the root of the *right subtree*.

We now show how a binary rooted tree can be used to represent the way in which an algebraic expression is evaluated.

Suppose we have an algebraic expression containing a number of variables and the binary operations of addition (+), subtraction (−), multiplication (×) and division (/). An example of such an expression is shown below:

$$(a - (b \,/\, c)) \times (d + e)$$

For the sake of clarity, we have used brackets explicitly to indicate the order in which the operations are to be carried out, including those around $b \,/\, c$ where brackets are not strictly necessary because of the convention in algebra about precedence of operators.

We now imagine that this expression is being evaluated by a computer with particular values substituted for the variables. The *principal operation* in this example is the multiplication, because this is the operation that would be executed *last*, after $(a - (b \,/\, c))$ and $(d + e)$ have both been evaluated. (This is exactly analogous to the principal connective of a logical expression, which we met in Chapter 4.) The principal operations of the subexpressions $(a - (b \,/\, c))$ and $(d + e)$ are the subtraction and the addition respectively. In turn, $(a - (b \,/\, c))$ contains the subexpression $b \,/\, c$ with division as its principal operation.

The structure of the expression can be depicted using a binary rooted tree known as an *expression tree* (Figure 11.8).

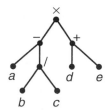

Figure 11.8

The root of the tree is labelled with the principal operation. The *left subexpression* $(a - (b \,/\, c))$ is represented by the left subtree, whose root is the left child of the root of the expression tree. Similarly, the *right subexpression* $(d + e)$ is represented by the right subtree, whose root is the right child of the root of the expression tree. The left subtree contains a right subtree of its own, corresponding to the subexpression $(b \,/\, c)$.

The way in which a computer processes the expression corresponds to visiting the vertices of the expression tree in a particular order. A procedure for visiting all of the vertices of a binary rooted tree in a given order is called a *traversal* of the tree. We will see how different ways of traversing an expression tree correspond to different ways of writing the expression.

The usual way of writing an algebraic expression is to place each operator (the symbol +, −, × or /) between the two expressions on which it operates: $3 + 2$, for example. This notation is called *infix* notation. Infix notation corresponds to a method of traversing the expression tree known as an *in-order traversal*.

The idea underlying in-order traversal is to visit all of the vertices corresponding to the left subexpression, then to visit the root, and lastly to visit the vertices corresponding to the right subexpression. The order in which the vertices are visited within each subexpression is itself an in-order traversal.

Here is an algorithm for carrying out an in-order traversal on a binary rooted tree T:

Algorithm *in-order_traverse*(T):
1. **If** T is not a leaf **then**
   1.1. *in-order_traverse*(left subtree of T)
2. Output the root of T
3. **If** T is not a leaf **then**
   3.1. *in-order_traverse*(right subtree of T)

Notice that the algorithm *in-order_traverse* contains calls to itself in Steps 1 and 3. This makes *in-order_traverse* a recursive algorithm.

Suppose we wish to apply the *in-order_traverse* algorithm to the expression tree in Figure 11.8. Here is a detailed description of what happens.

In Step 1, we are required to perform an in-order traversal of the left subtree with its root at the minus sign. To do this, we need to put execution of this first copy of the algorithm on hold while we begin executing a second copy with T equal to the left subtree. In Step 1 of this second copy, we need to go on hold again and begin execution of a third copy with T equal to the subtree with its root at a. This subtree is a leaf, so the third copy of the algorithm executes to completion and outputs a.

Returning now to the second copy, we execute Step 2 (output –) and move to Step 3. A third copy of the algorithm is needed again, this time with T equal to the subtree with its root at the division sign. Now a fourth copy is needed with T equal to the subtree with its root at b. This produces an output of b and returns to the third copy. The third copy continues execution and outputs /. Another fourth copy, this time using c as the root, produces an output of c. Execution of the second copy is now complete.

Step 2 of the first copy now executes, giving an output of ×.

We now move to Step 3 of the first copy. This requires us to perform an in-order traversal of the subtree with root at the plus sign. Calling up a second copy of the algorithm again, followed by a third copy, we output d, then return to the second copy to output +, then call up another third copy to output e. This completes the entire algorithm.

The output of the algorithm is the original expression with the variables and operators in the order written (but without the brackets).

All of this sounds much more complicated than it really is. Describing each step in detail requires a lot of words, but the process is actually fairly straightforward.

Two other traversal methods are *pre-order* and *post-order* traversal. The idea underlying pre-order traversal is to visit the root before visiting each subtree. A recursive algorithm for pre-order traversal is given below:

Algorithm *pre-order_traverse*(T):
1. Output the root of T
2. **If** T is not a leaf **then**
   2.1. *pre-order_traverse*(left subtree of T)
   2.2. *pre-order_traverse*(right subtree of T)

EXAMPLE 11.4.1

Carry out a pre-order traversal on the expression tree in Figure 11.8.

SOLUTION

Output the root: ×
   Pre-order traverse the left subtree. To do this, output its root (−), then pre-order traverse its left subtree (output: *a*), then pre-order traverse its right subtree (output: */ b c*).
   Pre-order traverse the right subtree, giving an output of + *d e*.
   The overall output is:

$$\times - a / b c + d e$$

An expression written in the form given by the output of a pre-order traversal is said to be written in *Polish prefix notation*. In Polish prefix notation, each operator appears before the two operands to which it is to be applied. Unlike infix notation, Polish prefix notation does not require the use of brackets to specify the order of the operations. Polish prefix notation is used in Lisp and related languages such as PC-Scheme.
   Finally, we look at post-order traversal. Here, the idea is to visit the root after visiting each subtree. A recursive algorithm for post-order traversal is given below:

Algorithm *post-order_traverse*(T):
1. **If** T is not a leaf **then**
   1.1. *post-order_traverse*(left subtree of T)
   1.2. *post-order_traverse*(right subtree of T)
2. Output the root of T

EXAMPLE 11.4.2

Carry out a post-order traversal on the expression tree in Figure 11.8.

SOLUTION

The details are similar to those in the previous example, so we just give the output:

$$a\,b\,c\,/-d\,e+\times$$

An expression written in the form given by the output of a post-order traversal is said to be written in *reverse Polish notation*. In reverse Polish notation, each operator appears after the two operands to which it is to be applied. Like Polish prefix notation, reverse Polish notation does not require the use of brackets to specify the order of the operations. Some calculators use reverse Polish notation — the numbers to be operated on are entered first (usually with the use of a key labelled ENTER), followed by the operation (e.g. addition).

Exercises

1. List the cycles in the graphs shown in Figure 11.9.

(a)
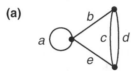

(b)
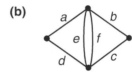

Figure 11.9

2. Draw all six trees (up to isomorphism) with six vertices.

3. Draw a tree whose vertices have the following degrees, or explain why no such tree exists:
   (a) seven vertices, with degrees 1, 1, 1, 1, 1, 3, 4;
   (b) eight vertices, with degrees 1, 1, 2, 2, 2, 2, 3, 3.

4. A tree has eight vertices of degree 1, three vertices of degree 2, and two vertices of degree 3. The other vertices all have degree 4. How many vertices must there be altogether?

5. Find a minimal spanning tree for each of the weighted graphs in Figure 11.10.

(a)
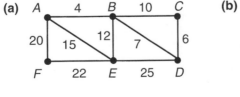

(b)
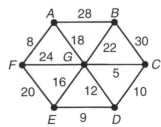

Figure 11.10

6. Write down the weight matrix for each of the weighted graphs in Exercise 5.

7. The relative costs of building the links between six nodes, denoted by *A*, *B*, *C*, *D*, *E* and *F*, in a proposed local area network are tabulated below:

	A	B	C	D	E
B	85				
C	–	95			
D	65	50	–		
E	55	–	–	70	
F	45	–	80	–	–

Find the network with the least possible cost that could be built to link all the nodes in the network.

8. Use Dijkstra's algorithm to find:

(a) the minimum distance path from *C* to *F* in the weighted graph in Exercise 5(a) (Figure 11.10(a));

(b) the minimum distance path from *A* to *C* in the weighted graph in Exercise 5(b) (Figure 11.10(b)).

9. Construct an expression tree for each of the following expressions:

(a) $a + ((b - c) \times d)$

(b) $((a \times b) / (c \times d)) - (e / f)$

(c) $(a - (b + (c + d))) - ((e \times f) \times g)$

10. By applying a pre-order traversal to the expression tree, write each of the expressions in Exercise 9 in Polish prefix notation.

11. By applying a post-order traversal to the expression tree, write each of the expressions in Exercise 9 in reverse Polish notation.

12. Write down (in infix notation) the expressions corresponding to the trees shown in Figure 11.11.

(a)   **(b)**

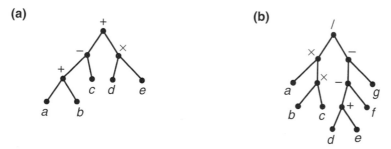

Figure 11.11

13. The binary rooted tree shown in Figure 11.12 can be used to encode and decode English text according to a *Huffman code*. The sequence of edges from the root to any letter yields the binary code for that letter. Note that the number of bits varies from one letter to another.

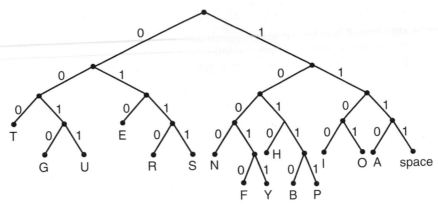

Figure 11.12

(a) Decode 00100110 11101011 11010111 10001010 01011010 11010011 11111100 01111111 10010001 11000.

(The coded message is grouped into strings of eight bits to enhance readability only; the grouping has no other significance.)

(b) Encode 'TO BE OR NOT TO BE'.

(c) What is the advantage of a Huffman code over codes that use a fixed number of bits for each letter?

12. NUMBER THEORY

12.1 *What is number theory?*

No doubt you have been familiar with the natural numbers since childhood. For this reason, it might seem that there could be little more to learn about them, at least in comparison with other types of numbers: fractions, negative numbers, irrational numbers and complex numbers. Nothing could be further from the truth; in fact, some of the most challenging problems in mathematics involve just the natural numbers. Since ancient times, mathematicians have been intrigued by the subtle properties that underlie the apparent simplicity of these numbers.

Number theory is principally the study of the natural numbers. However, the term is often extended to include the study of all the integers or all the rational numbers, since many of the techniques developed in the study of the natural numbers can also be applied to the study of these more general types of numbers. Even the study of irrational numbers can be included, because any problem that involves distinguishing between rational and irrational numbers (for example a proof that a certain number or class of numbers is irrational) invariably requires the use of properties of the natural numbers in order to solve it.

For our purposes, the main reason for studying number theory lies in its usefulness in many areas of computing. Until quite recently, number theory was viewed as a branch of mathematics with limited practical use, but the situation has changed dramatically as a result of the computer revolution. Data encryption, for example, relies heavily on results from number theory, as do the design of error-detecting and error-correcting codes and the generation of pseudo-random numbers for simulation programs, to name just some of the areas where number theory can be applied. It is not really surprising that the natural numbers, the most fundamental discrete structure in mathematics, should play a prominent role in the study of machines that process information in a discrete fashion.

12.2 *Divisibility and prime numbers*

As usual, we begin with some basic definitions. For much of the time we will be working with the integers (positive, zero and negative), but there will be some occasions when we will need to restrict our attention just to the natural numbers.

215

Definition

Let *a* and *b* be integers. We say *a* *divides* *b* (or *a* is a *divisor* of *b*) if there is an integer *n* such that $an = b$. We also say in this case that *b* is *divisible* by *a*, or that *b* is a *multiple* of *a*.

If *a* divides *b* we will write this as $a \mid b$. If *a* does not divide *b* then we write $a \nmid b$.

Informally, the definition says that *a* divides *b* if *a* 'goes into' *b* an exact number of times. For example, $4 \mid 12$ (because $4 \times 3 = 12$), but $5 \nmid 12$ (because there is no integer *n* such that $5n = 12$).

The following basic properties of divisibility are reasonably obvious after a moment's thought, and in any case they are not difficult to prove:

1. For any integer *a*, $a \mid 0$, $1 \mid a$ and $a \mid a$.
2. For any integers *a*, *b* and *c*, if $a \mid b$ then $a \mid bc$.
3. For any integers *a*, *b* and *c*, if $a \mid b$ and $a \mid c$ then $a \mid b + c$.
4. For any integers *a*, *b* and *c*, if $a \mid b$ and $b \mid c$ then $a \mid c$.
5. For any integers *a* and *b*, if $a \mid b$ then $-a \mid b$.
6. For any integers *a* and *b*, if $a \mid b$ and $b \mid a$ then either $a = b$ or $a = -b$.

As an example, we will prove the second property. Let $a \mid b$; this means that there is an integer *n* such that $an = b$. In order to show now that $a \mid bc$ (where *c* is any integer), we need to show that there is an integer *m* such that $am = bc$. There is indeed such an integer — it is *nc*. This concludes the proof.

Definition

A natural number *p* other than 1 is *prime* if its only positive divisors are 1 and *p*. A natural number other than 1 is *composite* if it is not prime.

Thus a composite number is a natural number that can be written as the product of smaller natural numbers. Note that 1 is neither prime nor composite.

The first few prime numbers are 2, 3, 5, 7, 11, 13, 17, 19, 23, ... The prime numbers are spaced irregularly, but tend to be distributed more sparsely among the natural numbers as they increase, according to a definite mathematical relationship (the 'Prime number theorem', which is well beyond the scope of this book).

By definition, any composite number can be expressed as the product of smaller natural numbers. If any of these smaller numbers are composite, they can in turn be factorised into yet smaller numbers, and so on. This process must eventually stop with an expression for the original number as a product of prime numbers. An expression of this form is called a *prime factorisation* of the original number.

For example, consider the composite number 72. One possible factorisation of 72 is 12×6. The numbers 12 and 6 can in turn be factorised as 4×3 and 2×3 respectively, and 4 can then be factorised as 2×2. It follows that a prime factorisation of 72 is $72 = 2 \times 2 \times 3 \times 2 \times 3$, which we can write more conveniently as $2^3 \times 3^2$.

If we think of each prime number as a 'product' of one prime (namely itself), then every natural number greater than 1 can be expressed as a product of prime numbers. The Fundamental theorem of arithmetic states that this prime factorisation can be done in essentially only one way.

Fundamental theorem of arithmetic

The prime factorisation of a natural number greater than 1 is unique, apart from the order of the factors.

Thus the prime factorisation $72 = 2^3 \times 3^2$ is the only prime factorisation of 72 (except that the three 2s and the two 3s could be written in a different order). Note that the factorisation would not be unique if we had defined 1 to be a prime number, because there would then be other factorisations, such as $1 \times 2^3 \times 3^2$ and $1^2 \times 2^3 \times 3^2$. It is mainly for this reason that prime numbers are defined in a way that explicitly excludes 1 from being a prime number.

If you think the Fundamental theorem of arithmetic is obvious, it is worthwhile digressing for a moment to look at a situation where a very similar statement turns out to be false. Imagine for a moment that we have decided to ban odd numbers and to work only with the positive even numbers: 2, 4, 6, 8, ... We can still add and multiply, because any sum or product of even numbers is an even number. Any even number that can't be factorised into smaller even numbers is 'prime' in this system; for example, 4 is still composite because $4 = 2 \times 2$, but 6 is 'prime' because it cannot be expressed as the product of even numbers. Now, 36 has two 'prime' factorisations: $36 = 2 \times 18 = 6 \times 6$, so the 'prime' factorisation of a number is not always unique in this system. This example should help to convince you that the Fundamental theorem of arithmetic is not a trivial result.

As far as the remainder of this chapter is concerned, we can take the Fundamental theorem of arithmetic as given. For the sake of completeness, however, we provide a proof in Section 12.6.

It has been known since ancient times that the number of primes is infinite. The proof uses the fact that if $a > 1$ and $a \mid b$ then $a \nmid b + 1$. If there were a finite number of primes, say n of them, we could (at least in principle) list them all: $p_1, p_2, p_3, \ldots, p_n$. The product $p_1 p_2 p_3 \ldots p_n$ of these primes would then have every prime as a divisor, and so $p_1 p_2 p_3 \ldots p_n + 1$ would have none of the primes as divisors. The prime factorisation of $p_1p_2p_3...p_n + 1$ would therefore involve new primes not in the original list, contrary to the assumption that the list contained all the primes. It follows that the number of primes must be infinite.

It might appear from this proof that new primes could be found by multiplying together all the known primes, adding 1, and finding the prime factorisation of the result. This method is not computationally feasible in practice, however, because of the difficulty of factorising very large numbers. The methods currently used to find new primes involve a combination of advanced mathematical theory and substantial computations performed on a computer.

Just for interest, we record here the largest known prime number at the time of writing — it is the 227832-digit number $2^{756839} - 1$.

12.3 *Greatest common divisors and the Euclidean algorithm*

In this section we will see how the greatest common divisor of two integers can be found using one of the oldest algorithms still in use.

Definitions

Let a and b be natural numbers.

The *greatest common divisor* of a and b is the largest natural number m such that $m \mid a$ and $m \mid b$.

The *least common multiple* of a and b is the smallest natural number m such that $a \mid m$ and $b \mid m$.

Two natural numbers a and b are *coprime* (or *relatively prime*) if their greatest common divisor is 1.

The greatest common divisor and least common multiple of a and b are denoted respectively by $\gcd(a, b)$ and $\mathrm{lcm}(a, b)$.

For example: $\gcd(27, 45) = 9$, $\gcd(15, 32) = 1$, $\mathrm{lcm}(12, 18) = 36$, $\mathrm{lcm}(11, 18) = 198$. The numbers 15 and 32 are coprime.

Both the greatest common divisor (gcd) and the least common multiple (lcm) arise when we perform arithmetic with fractions. The gcd arises when a fraction is to be reduced to its lowest terms; for example, $^{27}/_{45}$ can be simplified to $^3/_5$ by dividing the numerator and the denominator by 9, the gcd of 27 and 45. The lcm arises as the 'lowest common denominator' when fractions are added; for example, when adding $^5/_{12}$ and $^7/_{18}$, the fractions must first be expressed in terms of the common denominator 36, which is the lcm of 12 and 18.

If a and b are fairly small, $\gcd(a, b)$ can be found by listing all the divisors of a and b and looking for the largest number common to both lists. If the numbers are large, however, this method is cumbersome, and a more efficient method is needed.

For example, suppose we want to evaluate $\gcd(2737, 1225)$. It would be tedious to do this by listing all the divisors of the two numbers. We can speed up the process, however, by using a property of the gcd that we will now establish:

For any integers a, b and k, $\gcd(a, b) = \gcd(b, a - kb)$.

This result will follow immediately, if we can show that not just the *greatest* common divisors, but all the common divisors, are the same for both the pair (a, b) and the pair $(b, a - kb)$. It is not difficult to see that any common divisor of a and b must also be a divisor of $a - kb$ (and therefore must be a common divisor of b and $a - kb$). Conversely, since $a = kb + (a - kb)$, any common divisor of b and $a - kb$ must be a divisor of a (and therefore must be a common divisor of a and b). Hence all the common divisors are the same for the two pairs of numbers; in particular, the greatest common divisor must be the same.

This result is important because it allows a gcd problem involving large numbers to be reduced to a gcd problem with smaller numbers. Returning to the problem of evaluating $\gcd(2737, 1225)$, we can put $a = 2737$ and $b = 1225$, and choose k in such a way that $a - kb$ is as small as possible (but still non-negative). The value of k which achieves this is a div b (using the Pascal-style notation we introduced in Chapter 2 for truncated division in which the remainder is ignored). In this example, $k = 2737$ div $1225 = 2$.

The calculation now goes like this:

$$\gcd(2737, 1225) = \gcd(1225, 2737 - 2 \times 1225) = \gcd(1225, 287)$$

The process can be repeated with the new gcd. Divide 287 into 1225, obtaining a quotient of 4 and a remainder of 77:

$$\gcd(1225, 287) = \gcd(287, 1225 - 4 \times 287) = \gcd(287, 77)$$

Repeating the process five more times, we obtain:

$$\begin{aligned}
\gcd(287, 77) &= \gcd(77, 56) \\
&= \gcd(56, 21) \\
&= \gcd(21, 14) \\
&= \gcd(14, 7) \\
&= \gcd(7, 0) \\
&= 7
\end{aligned}$$

The procedure we have just described is known as the *Euclidean algorithm*.

If we regard gcd as a function with domain $\mathbf{N} \times \mathbf{N}$ and codomain $\mathbf{N}$, then the Euclidean algorithm can be written very succinctly in the following recursive form:

Algorithm gcd(a, b):
1. **If** $b = 0$ **then**
   1.1. gcd $\leftarrow a$
   else
   1.2. $k \leftarrow a$ div b
   1.3. gcd $\leftarrow$ gcd(b, $a - kb$)

A non-recursive algorithm for inputting two natural numbers and calculating their gcd using the Euclidean algorithm can be written using a **While-do** loop:

1. Input a, b
2. **While** $b \neq 0$ **do**
   2.1. $k \leftarrow a$ div b
   2.2. $b \leftarrow a - kb$
   2.3. $a \leftarrow b$
3. Output a

For the purpose of hand calculation, the computation is usually set out by writing down a set of equations, as shown in the following example.

EXAMPLE 12.3.1

| Use the Euclidean algorithm to evaluate gcd(2093, 836).

SOLUTION

$$2093 = 2 \times 836 + 421$$
$$836 = 1 \times 421 + 415$$
$$421 = 1 \times 415 + 6$$
$$415 = 69 \times 6 + 1$$
$$6 = 6 \times 1 + 0$$

Therefore gcd(2093, 836) = 1.

At each step of the calculation, the divisor and the remainder become respectively the dividend (number to be divided) and the divisor in the next step. The process terminates when the remainder is zero. The last non-zero remainder is the gcd.

The Euclidean algorithm is used in computer algebra software, which performs exact arithmetic with rational numbers in which the numerators and denominators can be arbitrarily large integers. An expression for a rational number as a ratio of two integers can be reduced to its lowest terms by dividing the numerator and the denominator by their gcd.

The Euclidean algorithm can also be used to find integer solutions x and y

to equations of the form $ax + by = c$, where the integers a, b and c are given. We will illustrate this by using the equations in Example 12.3.1 to find integers x and y such that $2093x + 836y = 1$. $\left(\text{If it were not for the restriction to integer}\right.$

solutions, we could of course choose x to be any number and set $y = \dfrac{1 - 2093x}{836}$. $\left.\right)$

Begin by rearranging the equations (except the last) to make the remainder the subject of the equation in each case:

$$421 = 2093 - 2 \times 836 \qquad (1)$$

$$415 = 836 - 1 \times 421 \qquad (2)$$

$$6 = 421 - 1 \times 415 \qquad (3)$$

$$1 = 415 - 69 \times 6 \qquad (4)$$

Starting with Equation 4, replace the divisor, 6, with the expression in Equation 3:

$$1 = 415 - 69 \, (421 - 1 \times 415)$$
$$= 70 \times 415 - 69 \times 421$$

Now replace 415 by the expression in Equation 2:

$$1 = 70 \, (836 - 1 \times 421) - 69 \times 421$$
$$= 70 \times 836 - 139 \times 421$$

Finally, replace 421 by the expression in Equation 1:

$$1 = 70 \times 836 - 139 \, (2093 - 2 \times 836)$$
$$= 348 \times 836 - 139 \times 2093$$

We can now see that $x = -139$, $y = 348$ is a solution of the equation $2093x + 836y = 1$.

EXAMPLE 12.3.2

Find a solution in integers to each of the following equations, or explain why no solution exists:

(a) $3024x + 2076y = 12$
(b) $3024x + 2076y = 36$
(c) $3024x + 2076y = 10$

SOLUTION

(a) The Euclidean algorithm applied to 3024 and 2076 gives the following equations:

$$3024 = 1 \times 2076 + 948$$
$$2076 = 2 \times 948 + 180$$
$$948 = 5 \times 180 + 48$$
$$180 = 3 \times 48 + 36$$
$$48 = 1 \times 36 + 12$$
$$36 = 3 \times 12 + 0$$

Rewriting the equations in terms of the remainders, we obtain:

$$948 = 3024 - 1 \times 2076$$
$$180 = 2076 - 2 \times 948$$
$$48 = 948 - 5 \times 180$$
$$36 = 180 - 3 \times 48$$
$$12 = 48 - 1 \times 36$$

Starting with the last of these and working back:

$$
\begin{aligned}
12 &= 48 - 1\,(180 - 3 \times 48) \\
&= 4 \times 48 - 1 \times 180 \\
&= 4\,(948 - 5 \times 180) - 1 \times 180 \\
&= 4 \times 948 - 21 \times 180 \\
&= 4 \times 948 - 21\,(2076 - 2 \times 948) \\
&= 46 \times 948 - 21 \times 2076 \\
&= 46\,(3024 - 1 \times 2076) - 21 \times 2076 \\
&= 46 \times 3024 - 67 \times 2076
\end{aligned}
$$

Therefore $x = 46$, $y = -67$ is a solution of the equation $3024x + 2076y = 12$.

(b) Since $36 = 12 \times 3$, a solution in integers to $3024x + 2076y = 36$ can be obtained from the solution to part (a) by multiplying both x and y by 3:

$$x = 3 \times 46 = 138, \; y = 3 \times (-67) = -201$$

(c) The equation has no solution in integers, because if x and y are integers then the left-hand side is divisible by 12 (12 is the gcd of 3024 and 2076) but the right-hand side is not.

Example 12.3.2 illustrates the three cases that arise in the general problem of finding integer solutions to equations of the form $ax + by = c$:

1. If $c = \gcd(a, b)$, then a solution is obtained directly from the Euclidean algorithm.
2. If c is a multiple of $\gcd(a, b)$, say $c = m \gcd(a, b)$, then a solution is obtained by multiplying the values of x and y obtained from the Euclidean algorithm by m.
3. If c is not a multiple of $\gcd(a, b)$, then the equation has no integer solutions.

The solution obtained in either Case 1 or Case 2 is not unique. When one solution, say $x = x_0$, $y = y_0$, has been found using the Euclidean algorithm, any pair of numbers of the form $x = x_0 + nb$, $y = y_0 - na$, where n is an integer, is also a solution, as you can check by substituting $x = x_0 + nb$ and $y = y_0 - na$ into the equation $ax + by = c$. In fact, it can be proved that all the integer solutions of the equation can be expressed in this form.

EXAMPLE **12.3.3**

| Find all the integer solutions to the equations in Example 12.3.2 (a) and (b).

SOLUTION

| The solutions to $3024x + 2076y = 12$ are $x = 46 + 2076n$, $y = -67 - 3024n$, where $n \in \mathbf{J}$.
| The solutions to $3024x + 2076y = 36$ are $x = 138 + 2076n$, $y = -201 - 3024n$, where $n \in \mathbf{J}$.

We have said very little about least common multiples apart from stating the definition, because we will not be needing them in what follows. For the sake of completeness, we state without proof the following result, which allows us to calculate the least common multiple of two natural numbers when we have found their greatest common divisor:

For any two natural numbers a and b, $\operatorname{lcm}(a, b) = \dfrac{ab}{\gcd(a, b)}$.

For example, $\operatorname{lcm}(3024, 2076) = \dfrac{3024 \times 2076}{\gcd(3024, 2076)} = \dfrac{3024 \times 2076}{12} = 523\,152$.

12.4 *Congruences*

The concept of divisibility leads naturally to the study of congruences. Congruence arithmetic was pioneered by the German mathematician Carl Friedrich Gauss (1777–1855), and is used today in a broad range of computing applications.

When we perform congruence arithmetic, we are effectively working with a finite subset of the set of all integers. There is nothing particularly new about this idea; when we quote the time in hours according to a 12-hour clock, we use only the integers from 1 to 12, and we know that after reaching 12 we 'start over again' at 1. As we will see shortly, it is often useful to think of congruence arithmetic in terms of a clock face.

Another example of a system in which arithmetic is carried out with a finite set of integers is the computer representation of integers. In Chapter 3, we investigated a model computer in which only the integers from –8 to 7 could be represented. Congruence arithmetic provides a more elegant mathematical setting for such studies.

The basic definition in the study of congruence arithmetic is as follows.

Definition

Let m be a natural number. Two integers a and b are said to be *congruent modulo m* if $m \mid b - a$.

We use the notation $a \equiv b$ mod m to indicate that a is congruent to b modulo m. The number m is called the *modulus*. If a is not congruent to b modulo m, we write $a \not\equiv b$ mod m.

An equivalent definition is: $a \equiv b$ mod m if a and b leave the same remainder after division by m.

Here are some examples:

- $7 \equiv 4$ mod 3, because both 7 and 4 leave a remainder of 1 after division by 3 (or, equivalently, because $7 - 4$ is divisible by 3).
- $11 \equiv 19$ mod 4, because both 11 and 9 leave a remainder of 3 after division by 4.
- $35 \equiv 14$ mod 7, because both 35 and 14 leave a remainder of 0 after division by 7.

In Chapter 2, we used 'mod' in a different but closely related sense — a mod b means the integer in the range from 0 to $b - 1$ that is congruent to a modulo b.

Sometimes it is useful to describe congruences in terms of the theory of relations which we studied in Chapter 5. If we think of the modulus m as fixed, then we can define a relation R on the set of integers by the rule: aRb if $a \equiv b$ mod m. It is not difficult to check that this relation satisfies the following three properties:

- $a \equiv a$ mod m, for all integers a.
- If $a \equiv b$ mod m, then $b \equiv a$ mod m, for all integers a and b.
- If $a \equiv b$ mod m, and $b \equiv c$ mod m, then $a \equiv c$ mod m, for all integers a, b and c.

In the terminology we introduced in Chapter 5, the relation is reflexive, symmetric and transitive, and therefore it is an equivalence relation. We could have expected that this would be the case anyway, because we know that equivalence relations are relations that state that two elements are 'the same' in some sense; in the present situation, two integers are related if they leave the same remainder after division by m.

In Chapter 5, we proved that any equivalence relation gives rise to a set of equivalence classes, in which all the elements in each class are related to each other. What are the equivalence classes in the present case?

We have already answered this question for one value of m. The relation in Example 5.5.5 in Chapter 5 is actually the relation $a \equiv b \bmod 4$, and in that example we derived the four equivalence classes. We recall them now:

$$E(0) = \{..., -12, -8, -4, 0, 4, 8, 12, ...\}$$

$$E(1) = \{..., -11, -7, -3, 1, 5, 9, 13, ...\}$$

$$E(2) = \{..., -10, -6, -2, 2, 6, 10, 14, ...\}$$

$$E(3) = \{..., -9, -5, -1, 3, 7, 11, 15, ...\}$$

We can write $E(k)$ in predicate form as follows, where $k = 0, 1, 2, 3$:

$$E(k) = \{4n + k: n \in \mathbf{J}\}$$

Thus, $E(0)$ is the set of all multiples of 4, $E(1)$ is the set of all numbers that are 1 more than a multiple of 4, and so on.

In general, the equivalence classes modulo m are the m sets $E(k) = \{mn + k: n \in \mathbf{J}\}$, where $k \in \{0, 1, 2, ..., m - 1\}$.

The arithmetic of congruences is based on the following results:

- If $a \equiv b \bmod m$ and $c \equiv d \bmod m$, then $a + c \equiv b + d \bmod m$.
- If $a \equiv b \bmod m$ and $c \equiv d \bmod m$, then $a - c \equiv b - d \bmod m$.
- If $a \equiv b \bmod m$ and $c \equiv d \bmod m$, then $ac \equiv bd \bmod m$.

These results can all be proved from the definition of 'congruent modulo m' and the properties of divisibility. We will leave the proofs of the first two results as exercises, and provide a proof of the third.

Let $a \equiv b \bmod m$ and $c \equiv d \bmod m$. This means that $m \mid a - b$ and $m \mid c - d$. Using divisibility Property 2 (see Section 12.2), we deduce that $m \mid ac - bc$ and $m \mid bc - bd$. It follows now from divisibility Property 3 that $m \mid (ac - bc) + (bc - bd)$. Rewriting this as $m \mid ac - bd$, we see that $ac \equiv bd \bmod m$, as required.

What these three results tell us is that addition, subtraction and multiplication 'work' when we do arithmetic modulo m. If we add, subtract or multiply two congruences modulo m, the resulting congruence is also a true statement.

Another way of looking at congruence arithmetic is from the point of view of equivalence classes. Suppose we concentrate just on addition for the moment. We can rephrase the first of the three results listed above in the following way: if a and b are in the same equivalence class, and c and d are in the same equivalence class, then $a + c$ and $b + d$ are in the same equivalence class. What this means is that *the equivalence classes themselves can be added* using the rule $E(a) + E(b) = E(a + b)$. This rule makes sense because the result of the addition is unchanged if a and b are replaced by other elements of their respective equivalence classes.

An example will help to make this clear. Look again at the equivalence classes we listed for modulus 4. Choose any two of them, say $E(2)$ and $E(3)$. Now, if we add *any* element of $E(2)$ to *any* element of $E(3)$, we always obtain an element of $E(1)$. We can sum up this observation by writing $E(2) + E(3) = E(1)$.

The complete addition table for the four equivalence classes is shown in Table 12.1.

	$E(0)$	$E(1)$	$E(2)$	$E(3)$
$E(0)$	$E(0)$	$E(1)$	$E(2)$	$E(3)$
$E(1)$	$E(1)$	$E(2)$	$E(3)$	$E(0)$
$E(2)$	$E(2)$	$E(3)$	$E(0)$	$E(1)$
$E(3)$	$E(3)$	$E(0)$	$E(1)$	$E(2)$

Table 12.1

We can think of addition of the equivalence classes modulo 4 as addition around a cycle (or 'clock face'):

To evaluate $E(2) + E(3)$, start at 2 and take 3 steps clockwise to obtain the answer: $E(1)$.

The multiplication table for the equivalence classes modulo 4 is shown in Table 12.2.

	$E(0)$	$E(1)$	$E(2)$	$E(3)$
$E(0)$	$E(0)$	$E(0)$	$E(0)$	$E(0)$
$E(1)$	$E(0)$	$E(1)$	$E(2)$	$E(3)$
$E(2)$	$E(0)$	$E(2)$	$E(0)$	$E(2)$
$E(3)$	$E(0)$	$E(3)$	$E(2)$	$E(1)$

Table 12.2

For example, $E(3) \times E(3)$ can be evaluated in the following way: $E(3) \times E(3) = E(3 \times 3) = E(9) = E(1)$, where the last equality is true because $9 \equiv 1 \bmod 4$.

In general, there is no operation of division in congruence arithmetic. At first sight, it might appear that we could divide one equivalence class by another using the rule $E(a)/E(b) = E(a/b)$. The problem with trying to divide using this rule is that the answer we get will depend on which elements we choose for a and b. For example, suppose we try to evaluate $E(0)/E(2)$ in modulo 4 arithmetic. If we choose $a = 0$ and $b = 2$, we obtain $E(0)$ as the answer. On the other hand, if we choose $a = 4$ and $b = 2$, we obtain an answer of $E(2)$. It follows that we cannot define division in a consistent manner, at least in this example. We will explore the division problem more fully shortly.

Congruence arithmetic provides some further insight into some aspects of computer arithmetic. In Chapter 3, we described the computer representation of integers using an imaginary computer that stored the integers from −8 to 7 as strings of 4 bits. At the time, the manner in which the computer performed arithmetic with these representations may have seemed a little mysterious. Using the theory of congruences we have developed, we can now provide a fuller explanation of why the method works.

Look again at Table 3.1 in Chapter 3. We observed in that chapter that the representation of a non-negative integer n in the 4-bit computer is n itself (in its 4-bit binary representation), and the representation of a negative integer n is $n + 16$. In either case, the integer and its representation are congruent modulo 16.

Now recall how the 4-bit computer adds two integers. The addition is performed on the representations using ordinary (binary) arithmetic, except that if the result of the addition is 16 or more, the 16s bit is ignored. This process is really just addition modulo 16.

We now see that the answer obtained by the computer must be congruent modulo 16 to the correct answer to the addition. Therefore, when the answer obtained by the computer is converted to the number that it represents, the result must be the correct answer, provided that it falls within the allowed range.

From what we now know about congruences, subtraction and multiplication can also be performed modulo 16. Thus, subtraction and multiplication can also be done on the 4-bit computer by performing the operation on the representations in the usual way, and ignoring the 16s bit, just as for addition.

In summary, the imaginary computer we studied in Chapter 3 performs arithmetic with integers modulo 16. In a similar way, a real computer that stores integers as, say, 16-bit strings, would carry out integer arithmetic modulo 2^{16}.

The lack of the operation of division in modular arithmetic is worth taking a closer look at. Suppose that $ax \equiv ay \bmod m$. It is tempting to try to divide both sides by a (or 'cancel the as'), but we know that we cannot do this; for example, $2 \times 0 \equiv 2 \times 2 \bmod 4$, but $0 \not\equiv 2 \bmod 4$.

If $ab \equiv ac \bmod m$, then by the definition of modulus, $m \mid ab - ac$. In general, it does not follow that $m \mid b - c$, so we cannot conclude that $b \equiv c \bmod m$. Is there any situation where we *can* make that deduction?

Yes, there is such a situation. The deduction is valid if a and m are coprime, that is, if $\gcd(a, m) = 1$. If this condition is satisfied, then division by a modulo m can be carried out.

Here is a proof that if $m \mid ab - ac$ and $\gcd(a, m) = 1$, then $m \mid b - c$. Notice first that if $\gcd(a, m) = 1$, then we can use the Euclidean algorithm to find integers x and y such that $ax + my = 1$. Now, since $m \mid ab - ac$, it follows that $m \mid x(ab - ac)$. By taking out a as a common factor, we can rewrite this statement as $m \mid ax(b - c)$. Next, using the equation $ax + my = 1$, we can replace ax by $1 - my$, to obtain $m \mid (1 - my)(b - c)$. By expanding the first bracket, we can write this as $m \mid b - c - my(b - c)$. Finally, since $m \mid my(b - c)$, we deduce that $m \mid b - c$. Therefore $b \equiv c \bmod m$, as required.

If we are working with congruences modulo 4, for example, we cannot expect to be able to divide by 2 or 4, because these numbers are not coprime with 4. However, we should be able to divide by 3 without any problems, because $\gcd(3, 4) = 1$.

If the modulus m is a prime number, we find ourselves in a very fortunate situation, because every integer that is not a multiple of m is coprime with m. For example, if $m = 5$ then we can carry out division by 1, 2, 3 or 4 (but not by 5). Expressed in terms of the equivalence classes, we can divide any equivalence class by $E(1)$, $E(2)$, $E(3)$ or $E(4)$.

How can we use the rule $E(a)/E(b) = E(a/b)$ to divide $E(3)$ by $E(4)$ modulo 5? Of course, there is no such equivalence class as $E(3/4)$. What we can do, however, is to apply the rule with another member of $E(3)$ in place of 3. We notice that $8 \in E(3)$, so the division can be carried out as follows:

$$E(3)/E(4) = E(8)/E(4) = E(8/4) = E(2)$$

The entire division table for arithmetic modulo 5 can be calculated in a similar way (noting that we cannot divide by $E(0)$ because it is the set of all multiples of 5). The result is shown in Table 12.3, in which the dividends and the divisors are listed in the left column and the top row respectively.

	$E(1)$	$E(2)$	$E(3)$	$E(4)$
$E(0)$	$E(0)$	$E(0)$	$E(0)$	$E(0)$
$E(1)$	$E(1)$	$E(3)$	$E(2)$	$E(4)$
$E(2)$	$E(2)$	$E(1)$	$E(4)$	$E(3)$
$E(3)$	$E(3)$	$E(4)$	$E(1)$	$E(2)$
$E(4)$	$E(4)$	$E(2)$	$E(3)$	$E(1)$

Table 12.3

Division can be used to solve a congruence for a variable x, as the next example shows.

EXAMPLE 12.4.1

Solve for x the congruence $9x + 5 \equiv 10 \bmod 11$.

SOLUTION

Subtract 5 from both sides of the congruence:

$$9x \equiv 5 \bmod 11$$

Since $\gcd(9, 11) = 1$, division by 9 is valid. Before we can perform the division, we need to choose a suitable multiple of 11 to add to the right-hand side in order to obtain a multiple of 9. We choose 22:

$$9x \equiv 27 \bmod 11$$

Now divide both sides by 9:

$$x \equiv 3 \bmod 11$$

The solution consists of all integers of the form $11n + 3$, where $n \in \mathbf{J}$.

If the modulus m is a prime number, as it is in Example 12.4.1, then the congruence $ax \equiv b \bmod m$ always has a solution if $m \nmid a$. However, it will not always be a straightforward exercise to find a constant to add to both sides of the congruence so that the division can be carried out, as we did in Example 12.4.1. In problems for which there is no obvious choice of such a constant, the Euclidean algorithm can be used to obtain the solution, as the following example illustrates.

EXAMPLE 12.4.2

Solve for x the congruence $18x + 13 \equiv 6 \bmod 23$.

SOLUTION

Subtract 13 from both sides:

$$18x \equiv -7 \bmod 23$$

By the definition of 'modulus', $23 \mid 18x - (-7)$, so there is an integer n such that $23n = 18x + 7$. Rewriting this equation as $23n - 18x = 7$, we can solve it using the Euclidean algorithm:

$$23 = 1 \times 18 + 5$$

$$18 = 3 \times 5 + 3$$

$$5 = 1 \times 3 + 2$$

$$3 = 1 \times 2 + 1$$

$$2 = 2 \times 1$$

Therefore:

$$
\begin{aligned}
1 &= 3 - 1 \times 2 \\
&= 3 - 1\,(5 - 1 \times 3) \\
&= 2 \times 3 - 1 \times 5 \\
&= 2\,(18 - 3 \times 5) - 1 \times 5 \\
&= 2 \times 18 - 7 \times 5 \\
&= 2 \times 18 - 7\,(23 - 1 \times 18) \\
&= 9 \times 18 - 7 \times 23
\end{aligned}
$$

We obtain $n = -7 \times 7 = -49$ and $x = -9 \times 7 = -63$ as a solution of the equation $23n - 18x = 7$. Hence $x = -63$ is a solution of the congruence $18x + 13 \equiv 6 \bmod 23$. The smallest non-negative solution is $x = -63 + 3 \times 23 = 6$, and the general solution is $x \equiv 6 \bmod 23$.

12.5 *Pseudo-random number generation*

For some applications, it is necessary to generate numbers randomly from a set of numbers, in such a way that each number in the set has an equal chance of being generated. There are many examples of situations in which random numbers are used; they include software for simulating a random process such as the arrival of customers at a counter, the selection of a random sample of people from a population in order to conduct an opinion poll, and the generation of test input for a computer program.

We can imagine how we might generate random numbers if we needed only a small number of them. For example, by throwing a fair six-sided die repeatedly, we can generate a random sequence of numbers from the set $\{1, 2, 3, 4, 5, 6\}$. The sequence of numbers produced in this way will have no pattern, but in the long run we expect each of the six numbers to appear about equally often.

If many random numbers are required, it is natural to ask whether a computer could be programmed to perform the task. A computer is a *deterministic* device — from the input fed into it, it produces output that is in principle always totally predictable. Nevertheless, it is possible to program a computer to generate sequences of numbers that have many of the properties of number sequences produced by a truly random process. Numbers generated in this manner on a computer are called *pseudo-random* numbers.

The generation of pseudo-random numbers is a large and complex topic, and we can only give a brief description here of one of the methods in common use.

Most of the methods currently in use for generating pseudo-random numbers are based on the following idea. A sequence of numbers is generated, which we may write like this:

$$x_0, x_1, x_2, x_3, \ldots$$

The first number, x_0, is called the *seed*, and is specified by the user. For any given pseudo-random number generator, different seeds yield different sequences, while choosing the same seed will always yield the same sequence. In some applications, it may be desirable to specify the seed to be a function of the time shown on the system clock, in order to avoid any possibility of bias on the part of the user.

Starting with the seed, the sequence is generated using a recursive formula:

$$x_i = f(x_{i-1}), \ i = 1, 2, 3, \ldots$$

where *f* is some function. This equation simply says that there is a rule that is used to calculate each term in the sequence from the previous term.

The *linear congruential method* is a method for generating a pseudo-random sequence of integers from the set $\{0, 1, 2, \ldots, m-1\}$. The sequence is generated using a rule of the form:

$$x_i \equiv ax_{i-1} + c \bmod m$$

In this congruence, *a* and *c* are constants, and x_i is the unique element of $\{0, 1, 2, \ldots, m-1\}$ that satisfies the congruence. Using 'mod' in the sense in which we used it in Chapter 2, we can write the rule in the following way:

$$x_i = (ax_{i-1} + c) \bmod m$$

If $c = 0$, the method is called the *multiplicative congruential method*.

The choice of the constants *a* and *c* is critical if the sequence is to behave like a sequence of true random numbers. We explore this problem now.

Notice that as soon as a term appears in the sequence for a second time, the sequence enters an infinitely repeating cycle. This is bound to happen eventually, because the terms belong to a finite set with *m* elements. The best we can do is to choose *m* to be large, and to try to ensure that the sequence is as long as possible before it starts repeating. The more random numbers we need for an application, the larger *m* must be, because we can generate at most *m* pseudo-random numbers before the sequence enters a repeating cycle. (More sophisticated arguments can be used to show that the sequence will show significant departures from randomness long before *m* numbers have been generated. In practice, it is usually recommended that no more than about $\sqrt{l}$ numbers be generated, where *l* is the length of the cycle.)

A convenient value to choose for *m* is the number of unsigned integers available on the machine we are using. (An unsigned integer is a non-negative integer, represented by its usual binary representation with no sign bit.) In order to explore the method, it will be convenient to use a small value of *m* in our first few examples. In Examples 12.5.1 to 12.5.3, we will assume that $m = 16$, corresponding to a machine on which 4 bits are used to represent unsigned integers. The integers available to us are therefore 0, 1, 2, …, 15. At best, we can hope to generate 16 pseudo-random numbers before the sequence starts repeating.

If we succeed in generating all 16 possible numbers, the choice of the seed will not matter, because a different seed will produce the 16 numbers in the same cyclic order, but starting at a different place. We will use 1 as the seed in the following examples.

The main question we would like to answer is: how do the values of *a* and *c* affect the resulting sequence? The next three examples show what can happen with different values of *a* and *c*.

EXAMPLE 12.5.1

Use the linear congruential method to generate a sequence of pseudo-random numbers modulo 16 with $a = 4$ and $c = 3$.

SOLUTION

The recurrence relation is:

$$x_0 = 1, \ x_i = (4x_{i-1} + 3) \bmod 16$$

Calculating the terms $x_1, x_2, x_3, \ldots$ successively, we obtain the sequence:

$$1, 7, 15, 15, \ldots$$

The sequence consists entirely of 15s from this point onwards.

The choice of a and c in Example 12.5.1 turned out to be unsatisfactory. In the next example we try different values of a and c.

EXAMPLE 12.5.2

Use the linear congruential method to generate a sequence of pseudo-random numbers modulo 16 with $a = 3$ and $c = 7$.

SOLUTION

The recurrence relation is:

$$x_0 = 1, \ x_i = (3x_{i-1} + 7) \bmod 16$$

The sequence is:

$$1, 10, 5, 6, 9, 2, 13, 14, 1, \ldots$$

The sequence now begins to repeat itself.

The sequence in Example 12.5.2 goes to eight terms before it starts repeating. This is certainly an improvement on the previous example, but it still falls short of the ideal situation in which all 16 numbers are generated.

EXAMPLE 12.5.3

Use the linear congruential method to generate a sequence of pseudo-random numbers modulo 16 with $a = 5$ and $c = 11$.

SOLUTION

The recurrence relation is:

$$x_0 = 1, \ x_i = (5x_{i-1} + 11) \bmod 16$$

The sequence is:

1, 0, 11, 2, 5, 4, 15, 6, 9, 8, 3, 10, 13, 12, 7, 14, 1, ...

The sequence in Example 12.5.3 contains all of the integers from 0 to 15. It is important to understand that this does not mean that it has all of the properties we would want a pseudo-random number sequence to have. In order to show that, it would be necessary to apply various statistical tests (and it would be inappropriate to do so anyway when we have used such a small value of m).

The examples we have just seen should make it clear that choosing appropriate values of a and c is not a trivial task.

In one sense, our examples have been a little misleading. We have been placing a high priority on generating all of the numbers in the set as our criterion for a 'good' sequence of pseudo-random numbers. In reality, it is far more important to generate a sequence that passes statistical tests for randomness, even at the expense of generating only a proper subset of the available numbers. The pseudo-random number generator in the next example is more typical of those used in practice.

EXAMPLE 12.5.4

Generate 20 pseudo-random numbers using the following generator:

$$x_0 = 2187, \ x_i = 2187x_{i-1} \bmod 65536$$

(Note that $65536 = 2^{16}$, so a generator of this form would be a natural choice on a machine that represents unsigned integers as 16-bit strings.)

SOLUTION

The numbers are:

64377, 21171, 32561, 38811, 10537, 41283, 42849, 59819, 14297, 6867, 10385, 36539, 22409, 53091, 45761, 5835, 47161, 52979, 62961, 4571.

Notice that the generator in Example 12.5.4 uses the multiplicative congruential method. It will not generate all of the numbers in the set $\{0, 1, 2, ..., 65535\}$; in fact, it is not difficult to see that all of the numbers generated must be odd, and a more sophisticated argument shows that it generates only one-quarter of the elements of the set. This is not a serious drawback; in most practical applications, each of the numbers would be divided by the modulus m to produce a sequence of pseudo-random numbers that are approximately uniformly distributed between 0 and 1, and for this purpose the constants used in Example 12.5.4 are satisfactory.

EXAMPLE 12.5.5

Convert the numbers obtained in Example 12.5.4 to numbers from a uniform distribution between 0 and 1.

SOLUTION

The conversion is carried out by dividing each number by 65536. The resulting sequence is:

0.982315, 0.323044, 0.496841, 0.592209, 0.160782, 0.629929, 0.653824, 0.912766, 0.218155, 0.104782, 0.158463, 0.557541, 0.341934, 0.810104, 0.698257, 0.089035, 0.719620, 0.808395, 0.960709, 0.069748.

12.6 *Proof of the Fundamental theorem of arithmetic*

Recall that the Fundamental theorem of arithmetic states that the prime factorisation of a natural number greater than 1 is unique, apart from the order of the factors.

Suppose we could find a natural number, n, with more than one prime factorisation. Then we could write one of the prime factorisations of n as $p_1p_2...p_k$, but there would also be a prime number q in a different prime factorisation of n such that q was not equal to any of the primes $p_1, p_2, ..., p_k$.

Now, since $q \mid n$, we can write $n \equiv 0 \bmod q$, and therefore $p_1p_2...p_k \equiv 0 \bmod q$. We know that none of the ps are equal to q, nor can any of the ps equal a multiple of q either, because they are all prime numbers. Therefore, using our results from Section 12.4, all of the ps can be used as divisors modulo q.

Divide both sides of the congruence $p_1p_2...p_k \equiv 0 \bmod q$ in turn by $p_1, p_2, ..., p_k$. The result is $1 \equiv 0 \bmod q$, which is impossible. Therefore the prime factorisation of n must be unique. This concludes the proof.

> ## Exercises

1. Prove the divisibility properties listed in Section 12.2.

2. Let n be any natural number greater than 1. Explain why the numbers $n! + 2$, $n! + 3$, $n! + 4$, ..., $n! + n$ must all be composite. (This exercise shows that it is possible to find arbitrarily long sequences of consecutive composite numbers.)

3. A list of all the prime numbers less than or equal to n can be generated in the following way: Begin with an empty list of primes. For each number from 2 to n in turn, test whether it is divisible by any of the primes in the list, and append it to the list if it isn't.

   Write this procedure as an algorithm.

4. Another method for generating all the prime numbers less than or equal to n, called the *Sieve of Eratosthenes*, is carried out as follows: Begin with a list of all the natural numbers from 2 to n. In the first step, remove from the list all the multiples of 2 (the first number on the list) except 2 itself. In the second step, remove from the list all the multiples of 3 (the next number on the list) except 3 itself. Similarly, in the third step, remove the multiples of 5 (the next number remaining on the list), and so on.

   (a) Obtain a list of all the prime numbers less than 50 using this method.
   (b) Write the method as an algorithm. (Define an array of Boolean variables $prime(2)$, $prime(3)$, $prime(4)$, ..., $prime(n)$, and initialise them all to 'true'. A number i can be 'removed' from the list by setting $prime(i)$ to 'false'.)

5. If a natural number n is being tested for primality, explain why it is sufficient to test it for divisibility by the natural numbers from 2 to $\sqrt{n}$.

6. For the following pairs of numbers, use the Euclidean algorithm to evaluate the gcd, and hence also find the lcm:

   (a) 572, 297   (b) 1384, 1144   (c) 1076, 2985

7. With the aid of the Euclidean algorithm, express the following fractions in their lowest terms:

   (a) $\dfrac{10881}{2067}$ (b) $\dfrac{2623}{4171}$

8. Find all the integer solutions to each of the following equations, or explain why no solution exists:

   (a) $4571x + 5726y = 21$
   (b) $2783x + 2046y = 10$
   (c) $4002x + 2091y = 3$

9. (a) Use the Euclidean algorithm to find all the integer solutions of the equation $8x + 5y = 100$.
   (b) Using the result of part (a), list all the solutions of $8x + 5y = 100$ for which x and y are natural numbers.

10. Let m be a natural number, and let R be the relation defined on **J** by $a R b$ if $a \equiv b \bmod m$. Prove that R is an equivalence relation.

11. Write down the equivalence classes modulo 5 in enumerated and predicate form.

12. Construct the addition and multiplication tables for the equivalence classes modulo 5.

13. Prove:

   (a) If $a \equiv b \bmod m$ and $c \equiv d \bmod m$, then $a + c \equiv b + d \bmod m$.

   (b) If $a \equiv b \bmod m$ and $c \equiv d \bmod m$, then $a - c \equiv b - d \bmod m$.

14. Which equivalence classes can be used as divisors when arithmetic is performed (a) modulo 9; (b) modulo 10?

15. Solve the following congruences for x, expressing the answer in the form $x \equiv a \bmod m$, where a is non-negative and is chosen to be as small as possible:

   (a) $7x \equiv 24 \bmod 19$ **(b)** $35x + 14 \equiv 23 \bmod 27$

16. Use the linear congruential method to generate a sequence of pseudo-random numbers modulo 16 with $a = 7$ and $c = 11$, using a value of 1 for the seed.

17. Use the linear congruential method to generate a sequence of pseudo-random numbers modulo 16 with $a = 9$ and $c = 11$, using a value of 1 for the seed.

18. It can be shown that the following conditions are necessary and sufficient in order for the linear congruential method to generate a cycle consisting of all of the integers from 0 to $m - 1$:

   - c and m are coprime;
   - $a - 1$ is divisible by every prime divisor of m;
   - if $4 \mid m$, then $4 \mid a - 1$.

   Verify this result for the linear congruential generators in Exercises 16 and 17.

19. Suppose we were to construct a linear congruential pseudo-random number generator in which c and m are coprime and $a = 1$. Show that this generator satisfies the three conditions in Exercise 18. Explain why it is nevertheless unsatisfactory as a pseudo-random number generator.

20. With the aid of a statistical calculator, calculate the sample mean and sample variance of the numbers obtained in Example 12.5.5, and compare the results with the theoretical values of 0.5 and $\frac{1}{12}$ respectively for a uniform distribution between 0 and 1.

13. ALGORITHMS AND COMPUTATIONAL COMPLEXITY

13.1 *How long does an algorithm take to run?*

In Chapter 1 we introduced the concept of an algorithm. Our main purpose at that stage was to define a language, or pseudocode, in which algorithms could be written. We have been using the ideas of Chapter 1, and the pseudocode notation in particular, throughout this book.

In this final chapter, we return to the study of algorithms themselves. In particular, we are interested in the following practical question: how long does an algorithm take to run?

Imagine that we are running a program on a computer, with a set of test inputs. The time (in seconds, say) between when the program is submitted and when the output appears will depend on many factors. Naturally, we expect that it will depend on the power of the machine itself — a supercomputer will do the job faster than a microcomputer. The time will also depend on the programming language, the compiler, the operating system, the available memory, any auxiliary storage devices (such as disks) used by the program, and possibly on other factors as well. In a multi-user system, the time will increase in proportion to the number of users. Finally, we expect that the time taken will depend on the algorithm used and the input.

It is only these last two — the algorithm and the input — that we will consider in this chapter. If we were to try to take the other factors into account as well, we would need to investigate questions about computer systems that would go well beyond the scope of this book. So, our investigations will not allow us to calculate actual running times in seconds on a real computer. Nevertheless, we will be able to compare algorithms one with another to determine which algorithm would run faster. It will also be possible in some cases to establish that certain algorithms are intractable for practical purposes, even if a very powerful machine is available.

In order to analyse the running time of an algorithm (or its *time complexity*, to use the technical term) without becoming involved in very complicated calculations, it will be necessary to make a number of approximations. We will make four approximations as we investigate the problem, and each one will be stated explicitly as it arises.

13.2 *Dominant operations and the first two approximations*

In order to begin to analyse the time complexity of an algorithm, we need some idea of the relative computation times of the basic operations performed by a computer in the course of carrying out a computation. Here is a list of the main operations we used in the algorithms in Chapter 1:

- Assignment ($\leftarrow$)
- Comparison ($=, \neq, <, >, \leq, \geq$)
- Arithmetical operations ($+, -, \times, /$)
- Logical operations (and, or, not)

How long does each of these operations take to execute? We cannot give a precise answer to this question without further information, because the actual times, and even the time taken by one operation relative to another, can vary between machines. Nevertheless, we can say in general that assignments are very fast; comparisons, addition, subtraction and the logical operations are slower; and multiplications and divisions are the slowest of the operations listed.

Most algorithms will use several of these operations, and keeping track of the times taken by all of them would be difficult. In order to simplify matters, we introduce the first of the four approximations.

First approximation: *Only count the most time-consuming operations.*

In particular, if an algorithm contains a significant number of multiplications or divisions, we will count only the multiplications and divisions when analysing the time complexity, and ignore the other operations altogether. This will be the situation in most algorithms that involve substantial numerical calculations. In algorithms where little or no arithmetic is performed, other operations will be the most time-consuming; in a sorting algorithm, for example, it is the time taken to compare the objects being sorted that dominates the time complexity of the algorithm.

EXAMPLE 13.2.1

What is the dominant (i.e. the most time-consuming) operation in the algorithm in Example 1.2.1 for finding the smallest number in a list of numbers?

SOLUTION

For convenience, we repeat the algorithm here:

1. Input the number of values n
2. Input the list of numbers $x_1, x_2, ..., x_n$
3. $min \leftarrow x_1$
4. **For** $i = 2$ **to** n **do**
    4.1. **If** $x_i < min$ **then**
        4.1.1. $min \leftarrow x_i$
5. Output min

The algorithm contains assignments in Steps 3 and 4.1.1, and a comparison in Step 4.1. There is no arithmetic (at least not explicitly — the **For-do** involves incrementing the loop variable i with each pass through the loop, but a computer does this very quickly). The dominant operation is therefore the comparison in Step 4.1.

EXAMPLE 13.2.2

What is the dominant operation in the algorithm in Example 1.3.2 for evaluating x^n where n is a positive integer?

SOLUTION

Here is the algorithm again:

1. Input x, n
2. $answer \leftarrow x$
3. **For** $i = 1$ **to** $n - 1$ **do**
    3.1. $answer \leftarrow answer \times x$
4. Output $answer$

The multiplication in Step 3.1 is the dominant operation.

It should be obvious that we are interested mainly in operations that are executed many times, especially operations within loops. An addition that is executed many times inside a **For-do** loop has a greater influence on the computation time than a multiplication that executes just once.

When once the time-consuming operations have been identified, the next step is to determine how many times each operation is executed when the algorithm runs. The answer usually depends on the input. In the algorithm for finding the smallest number in a list, for example, the comparison in Step 4.1 is executed $n - 1$ times, where n is the length of the list. In the algorithm for evaluating x^n, the multiplication in Step 3.1 is executed $n - 1$ times.

If an algorithm contains a **While-do** or a **Repeat-until** structure, the problem is more complicated. The algorithm in Example 1.3.1 for checking whether a string of n characters consists entirely of digits illustrates this situation.

1. Input n
2. Input $c_1 c_2 \ldots c_n$
3. $i \leftarrow 0$; *nondigit_detected* $\leftarrow$ false
4. **While** *nondigit_detected* = false and $i < n$ **do**
   4.1. $i \leftarrow i + 1$
   4.2. **If** c_i is not a digit **then**
         4.2.1. *nondigit_detected* $\leftarrow$ true
5. **If** *nondigit_detected* = true **then**
   5.1. Output 'The string contains non-digit characters.'
   else
   5.2. Output 'The string consists entirely of digits.'

The main part of the computation takes place in the **While-do** in Step 4. The dominant operation is the test in Step 4.2 to find out if c_i is a digit, which essentially involves making a set of comparisons (checking c_i against each of the 10 digits in turn).

We now need to find out how many times the **While-do** loop is executed. If the input string consists entirely of digits then the loop is executed n times, but if a non-digit character is present then the number of executions is smaller. There is no way of determining the number of executions without running the algorithm (or counting the number of characters to the first non-zero character, which amounts to the same thing).

There are two approaches we could take here to get around the problem. One is to try to estimate the *average-case time complexity* — if we take the length of the string to be given, the time will differ from one string to another, but we could estimate how long it would take on average. The more common approach, and the one we will pursue here, is to deal with the *worst-case time complexity*. This leads us to our second approximation.

Second approximation: *Where the computation time can vary with a given input size, analyse the worst case.*

In the present example, the worst case (in the sense that it takes the longest time) occurs when there are no non-digit characters in the string, and the loop executes n times.

13.3 *Comparing functions and the third approximation*

One reason for wanting to estimate the time complexity of algorithms is to be able to compare two algorithms for accomplishing the same task, to find out which one runs faster. This sounds straightforward — estimate the time complexity of each algorithm and see which answer is smaller. However, as the examples in the previous section show, the time complexity of an algorithm is usually a function of the size of the input. (We haven't given a general definition of 'size of the input', but have used a natural measure of size in each of the examples we have considered.) One algorithm might be faster with some inputs, while another might be faster with other inputs. We need to find a way of dealing with this difficulty.

In order to explore the question further, it will be convenient to have a particular problem to work with as an example.

Suppose we would like to construct a faster algorithm for evaluating x^n (where n is a positive integer). In this case, 'faster' means the new algorithm would use fewer than $n - 1$ multiplications. Is such an algorithm possible? For instance, could we evaluate x^4 (for a given value of x) using fewer than three multiplications? Yes; x^2 can be evaluated using one multiplication, then the result can be squared (which involves a second multiplication) to obtain the answer, $(x^2)^2 = x^4$.

This idea can be generalised to permit the evaluation of x^n very quickly if n is any power of 2. For example, x^{16} can be evaluated using just four multiplications by computing $(((x^2)^2)^2)^2$, whereas the earlier algorithm requires 15 multiplications.

If n is not a power of 2, we can still exploit the idea of 'repeated squaring'. For example, if we want to evaluate x^{13}, we can begin by evaluating x^8 using the repeated squaring process. This leaves x^5 to be evaluated and then multiplied by x^8 to give x^{13}. Now x^5 can be evaluated by first evaluating x^4 (again by repeated squaring) and multiplying the answer by x.

In general, x^n can be evaluated by finding the highest power of 2 that is less than or equal to n (call it k), evaluating x^k by repeated squaring, evaluating x^{n-k}, and multiplying x^k by x^{n-k}. The evaluation of x^{n-k} would itself be performed using this procedure.

What we have just described is a recursive algorithm for evaluating x^n. From the point of view of computational complexity, however, we can do a little better in this problem if we avoid recursion. In the evaluation of x^{13}, it would be preferable to evaluate x^4 before x^8, because x^8 could then be obtained simply by squaring x^4. In the general problem of evaluating x^n, we could evaluate successively x^2, x^4, x^8, x^{16}, ... as far as needed, at each stage using the power of x if it is needed and ignoring it if it is not.

We need to establish a procedure for determining which of these powers are needed. For the case $n = 13$, we evaluate x^{13} as $x^8 x^4 x$, which we can write as $x^8 x^4 x^1$. Notice that the numbers 8, 4 and 1 are powers of 2 that add up to 13. The key to the algorithm is therefore to be able to express n as a sum of distinct powers of 2.

We have seen this before, but in a different guise. The binary representation of a number is essentially its representation as a sum of powers of 2. For example:

$$13 = 8 + 4 + 1$$
$$= 2^3 + 2^2 + 2^0$$
$$= 1 \times 2^3 + 1 \times 2^2 + 0 \times 2^1 + 1 \times 2^0$$
$$= 1101_2$$

Rather conveniently, n is already stored in binary form in a computer, so no explicit conversion to binary is needed as far as the computer implementation of the algorithm is concerned.

The algorithm in pseudocode is shown below. (In the algorithm, there is a superfluous multiplication by 1 the first time Step 4.2.1 is executed if $b_1 = 0$. Rewriting the algorithm to avoid the extra multiplication would make it more complicated, and the decrease in time complexity would be negligible.)

1. Input x, n
2. $d \leftarrow$ number of bits in n
{The binary representation of n is $b_d b_{d-1} b_{d-2} \ldots b_1$.}
3. **If** $b_1 = 1$ **then**
      3.1. *answer* $\leftarrow x$
     else
      3.2. *answer* $\leftarrow 1$
4. **For** $i = 2$ **to** d **do**
      4.1. $x \leftarrow x \times x$
      4.2. **If** $b_i = 1$ **then**
         4.2.1. *answer* $\leftarrow$ *answer* $\times x$
5. Output *answer*

If x_0 denotes the original value of x, then the successive values of x after Step 4.1 is executed are x_0^2, x_0^4, x_0^8, x_0^{16} … The variable *answer* is multiplied successively by each of these powers of x_0 for which the corresponding bit in the binary representation of n is 1. It is this algorithm, or some variation of it, that is most commonly used in practice for evaluating x^n for positive integer values of n on a computer.

How does the new algorithm compare with the earlier one in its time complexity? The dominant operation in both algorithms is multiplication, so we can carry out the comparison by counting the number of multiplications performed by each algorithm. Table 13.1 shows the numbers of multiplications performed by the two algorithms, for selected values of n.

As expected, the second algorithm usually performs better than the first, with the improvement becoming more marked for higher values of n. The one exception is $n = 2$, where the second algorithm is slightly worse. When we analyse the time complexity of an algorithm, we are generally only interested in what happens for large inputs. After all, any algorithm will surely be fast if

n	First algorithm	Second algorithm
1	0	0
2	1	2
3	2	2
4	3	3
5	4	3
6	5	4
7	6	4
8	7	4
9	8	4
10	9	5
11	10	5
12	11	5
13	12	5
14	13	6
15	14	6
16	15	5
32	31	6
64	63	7

Table 13.1

the input is small; it is only when the input is large that the running time is likely to be a cause for concern. This leads to the third approximation.

Third approximation: *Assume that the input is large.*

With this approximation in mind, it appears that the second algorithm for evaluating x^n is better than the first, although we have not yet proved that this is the case. In order to construct a proof, we will need a formula for the number of multiplications performed by the second algorithm as a function of the input size. In the worst case (remembering our second approximation), Step 4.2.1 will be executed with every pass through the **For-do** loop. The number of multiplications in the worst case is therefore $2(d-1)$, where d is the number of bits in the binary representation of n.

Up to this point, we have been rather imprecise in specifying what we mean by 'size of the input'. Now that we are faced with two different measures of size, n itself and the number of bits in the binary representation of n, we need to decide exactly how the size of the input should be measured. Since any input to a digital computer can be expressed as a string of bits, it is really the number of bits in the input that is the fundamental measure of input size. From now on, this is the measure of input size we will use. (The input in the present problem also includes x, but this can be ignored because we are assuming that the size of x does not affect the time complexity.)

The time complexity of the second algorithm is already expressed in terms of the number of bits, d; it remains for us to express the time complexity of the first algorithm in terms of d also.

Suppose that the input to the first algorithm is a string of d bits corresponding to the binary representation of n. If we regard d as fixed, the 'worst case' occurs when n is the largest number with d bits in its binary representation — $n = 2^d - 1$. We already know that the time complexity of the algorithm is $n - 1$. The time complexity in terms of d therefore equals $2^d - 2$.

In order to establish what we suspect to be the case, that the second algorithm is the faster one, it remains only to show that $2(d - 1)$ is less than $2^d - 2$ when d is sufficiently large. This can be done by proving that $2d < 2^d$ when $d \geq 3$. The proof, using induction on d with $d = 3$ as the base step, is left as an exercise.

In general, the time complexity of an algorithm is a function: $f: \mathbf{N} \to \mathbf{N}$, $f(n) = $ maximum number of dominant operations performed if the input size is n.

When once the time complexity has been expressed as a function in this way, comparing two algorithms amounts to comparing two functions with domain and codomain $\mathbf{N}$, to determine which function takes larger values when the variable is sufficiently large. For this purpose, it will pay us to investigate some of the functions that can arise as the time complexity of algorithms.

Although we have specified $\mathbf{N}$ as the codomain in our definition of the time complexity of an algorithm, it is actually more convenient to treat time complexity as a *real-valued* function. A function of this type may take non-integer values, and yet can still be a good approximation to the true time complexity. For the rest of this chapter, we will take $\mathbf{R}$ rather than $\mathbf{N}$ to be the codomain of the functions we will be studying.

Among the most important functions arising in this context are the *polynomial* functions:

$$f(n) = a_k n^k + a_{k-1} n^{k-1} + a_{k-2} n^{k-2} \ldots + a_0$$

where the constants $a_k, a_{k-1}, a_{k-2}, \ldots, a_0$ (the *coefficients*) are real numbers, and the non-negative integer k is the *degree* of the polynomial. The algorithms in Examples 13.2.1 and 13.2.2 both have time complexity[1] $n - 1$, which is a polynomial with degree 1. An algorithm with nested **For-do** loops, such as the one in Chapter 1, Exercise 12, may have a polynomial of degree 2 as its time complexity. In fact, if we take the dominant operation in that algorithm to be the divisibility test performed within the inner **For-do**, then the time complexity of that algorithm is n^2, where n is the natural number input.

---

[1]To be precise but pedantic, the time complexity is the function f whose rule is $f(n) = n - 1$. In practice, it is common to refer to the formula for $f(n)$ as the 'time complexity'.

EXAMPLE 13.3.1

Two algorithms, A and B, have time complexities f and g respectively, where $f(n) = n^2$ and $g(n) = 3n^2 - 3n + 1$. Determine which algorithm is faster.

SOLUTION

A good starting-point is to try to find a value of n such that $f(n) = g(n)$. We can do this here by solving the equation $n^2 = 3n^2 - 3n + 1$. Collecting all the terms on one side yields the quadratic equation:

$$2n^2 - 3n + 1 = 0$$

The left-hand side can now be factorised:

$$(2n - 1)(n - 1) = 0$$

The solutions are $n = 1/2$ and $n = 1$. The larger solution, $n = 1$, is the one of interest from the point of view of analysing the behaviour for large values of n. We expect that one of the two functions f and g will be larger for all values of n greater than 1. This is in fact the case; if $n > 1$ then the two factors, $2n - 1$ and $n - 1$, are both positive, so $2n^2 - 3n + 1 > 0$. Hence $3n^2 - 3n + 1 > n^2$, so Algorithm A is faster.

In general, it is the term of the polynomial that contains the highest power of n that is critical in the analysis of time complexity. For this reason, *monomials* (polynomials with only one term) are especially important in this work.

EXAMPLE 13.3.2

Two algorithms, A and B, have time complexities f and g respectively, where $f(n) = n^3$ and $g(n) = 10n^2$. Determine which algorithm is faster.

SOLUTION

We begin by solving the equation $n^3 = 10n^2$. Dividing both sides by n^2 (which is permissible, because we may assume in this application that n is positive), we obtain $n = 10$. It is now clear that $n^3 > 10n^2$ when $n > 10$ (just multiply both sides of the inequality $n > 10$ by the positive quantity n^2). Therefore $f(n) > g(n)$ when n is sufficiently large, so Algorithm B is faster.

Essentially the same argument as the one we used in Example 13.3.2 can be used to establish the following result: if $k > k'$ and if a and b are positive constants, then $an^k > bn^{k'}$ for sufficiently large values of n. It is the exponents k and k' that really matter; if they are different, then the values of a and b have no effect on the result, provided that n is sufficiently large.

13.4 *The fourth approximation and the $O(f)$ notation*

The three approximations we have made so far have certainly seemed very reasonable. By contrast, our fourth and last approximation, for which we will introduce the notation $O(f)$ (the so-called 'big O' notation), will probably appear rather drastic. Nevertheless, this approximation is a useful one, and is commonly made in the analysis of the time complexity of algorithms.

We will introduce the fourth approximation by considering the following sequence of functions of n, which we can think of as time complexities:

$$n, 2n, 3n, 4n, \ldots$$

We can see that each term is greater than the previous term, provided that n is positive. If the terms are the time complexities of algorithms, then the algorithms become slower as we move further along the sequence. In fact, we can state precisely by what factor any given term is greater than some earlier term; $6n$ is 3 times $2n$, for example.

Now consider the function n^2. It is greater than all of the functions in the sequence above, in the sense in which we have been comparing time complexities. (To see this, note that $n^2 > kn$ when $n > k$, no matter what the value of k is.) It is not possible to state a constant factor by which n^2 is greater than kn, however, because the ratio of n^2 to kn is n/k, which increases without bound as n increases. We say that n^2 is an 'order of magnitude' greater than the functions in the list above.

When we make the fourth of our approximations, all of the functions n, $2n$, $3n$, $4n$, $\ldots$ will be treated as the same, while n^2 will be treated as larger. It may seem surprising that such a drastic approximation could be useful; after all, it means, for example, that $1\ 000\ 000n$ is to be treated no differently from n itself, even though the first function is a million times larger than the second. It is certainly true that this approximation will have to be treated with caution, and there will be times when it will not be appropriate to make it. Nevertheless, it turns out to be very useful.

Fourth approximation: *Don't distinguish between two time complexities if the growth rate of one is a constant multiple of the growth rate of the other.*

In order to implement this approximation in practice, the following definition is useful.

Definition

Let f and g be time complexities. Then f is $O(g)$ if there is a positive number c such that $f(n) \leq cg(n)$ for all sufficiently large values of n. ($O(g)$ is read 'big O of g', where O is the letter 'oh', not the number zero.)

The expression 'sufficiently large' in the above definition means that there is a number N such that the condition $f(n) \leq cg(n)$ is satisfied whenever $n > N$ (but not necessarily when $n \leq N$). This is consistent with the third approximation, which says that we need only consider what happens when the input is large.

Informally speaking, 'f is $O(g)$' means f is less than or equal to g in the sense of the fourth approximation. If Algorithms A and B have time complexities f and g respectively, and f is $O(g)$, then either A is faster than B or A has a running time of the same order of magnitude as B.

When the 'big O' definition is used in practice, f is usually the time complexity of the algorithm we are investigating, and g is a simple function such as $g(n) = n^k$.

EXAMPLE 13.4.1

| Let $f(n) = 5n$ and $g(n) = n$. Show that f is $O(g)$.

SOLUTION

| In order to show that f is $O(g)$, we need to find a positive number c such that $5n \leq cn$ for all sufficiently large values of n. There is an obvious value of c that satisfies this requirement: $c = 5$. Therefore f is $O(g)$.

The value of c in Example 13.4.1 is not unique; we could have set c equal to 6, or indeed to any number greater than 5, and the reasoning would still have been correct.

EXAMPLE 13.4.2

| Let $f(n) = 3n + 10$ and $g(n) = n$. Show that f is $O(g)$.

SOLUTION

| We seek a positive number c such that $3n + 10 \leq cn$ when n is sufficiently large. It is clear that $c = 3$ won't work, so we try $c = 4$. There are some values of n ($n = 1$, for example) for which $3n + 10 > 4n$; however, it could still be the case that $3n + 10 \leq 4n$ when n is large, so we should check this possibility before we rule out $c = 4$. Solving the equation $3n + 10 = 4n$, we obtain $n = 10$ as the critical value of n.

| The reasoning now runs as follows: Let $n \geq 10$. Then $3n + n \geq 3n + 10$, so $3n + 10 \leq 4n$. Therefore the condition is satisfied with $c = 4$, so f is $O(g)$.

In practice, only the final steps of the reasoning should be presented as the solution; the rest is 'rough working'.

EXAMPLE 13.4.3

| Let $f(n) = n^2$ and $g(n) = n$. Show that f is not $O(g)$.

SOLUTION

In order to show that f is not $O(g)$, we have to prove that there is no value of c satisfying the condition in the definition. In other words, we need to show that there is no number c such that $n^2 \leq cn$ for all sufficiently large values of n.

Suppose that such a number c did exist. Then from $n^2 \leq cn$ it would follow that $n \leq c$, for all sufficiently large values of n. But this is clearly impossible; no matter what value we give to c, n could always be chosen to be a larger number. Therefore f is not $O(g)$.

The next example is an important one, because it illustrates a technique that can be applied whenever the time complexity of an algorithm is a polynomial function.

EXAMPLE 13.4.4

| Let $f(n) = 2n^4 - 5n^3 + 10n^2 - 6n + 9$ and $g(n) = n^4$. Show that f is $O(g)$.

SOLUTION

The definition of 'big O' refers to 'sufficiently large' values of n, so we may assume that n is positive. If the negative terms of the polynomial are deleted, the resulting polynomial will be larger. Thus:

$$2n^4 - 5n^3 + 10n^2 - 6n + 9 \leq 2n^4 + 10n^2 + 9$$

If some of the terms on the right-hand side are replaced by terms with a higher power of n, the new terms will be larger. Thus we can write:

$$2n^4 + 10n^2 + 9 \leq 2n^4 + 10n^4 + 9n^4$$

Putting the two inequalities together, we obtain:

$$2n^4 - 5n^3 + 10n^2 - 6n + 9 \leq 21n^4$$

Taking $c = 21$ in the definition, we conclude that f is $O(g)$.

There is no need to find the smallest value of c that works. In Example 13.4.4, it would be possible, with more effort, to find a value of c smaller than 21 that could be used, but there would be no point in doing so.

The technique used in Example 13.4.4 can be used to establish the following important result.

Theorem

If f is a polynomial with degree k, then[2] f is $O(n^k)$.

For example, if $f(n) = 6n^2 + 8n - 5$ then f is $O(n^2)$. It would also be true to say that f is $O(n^3)$, or indeed that f is $O(n^k)$ where k is any number greater than 2, but these statements are weaker than the statement that f is $O(n^2)$; they don't say as much as 'f is $O(n^2)$' is saying. As a general rule, we try to make the strongest statement we possibly can; hence we say in this case that f is $O(n^2)$ rather than, for instance, f is $O(n^3)$.

How does all this work in practice? If two algorithms have time complexity $O(n^2)$, then their running times have the same 'order of magnitude', although it could still be the case that one is many times faster than the other. If a third algorithm has time complexity $O(n^3)$, then it is slower than the other two by an order of magnitude. Not only would it run more slowly than the other two algorithms (for large input), but the discrepancy in the running times would increase rapidly as the size of the input increased. Provided we exercise a little caution, we can usually say that an algorithm with time complexity $O(n^2)$ will be faster in practice than one with time complexity $O(n^3)$.

13.5 *Sorting algorithms*

In order to put the theory we have been developing into a practical setting, we will look at the problem of designing an algorithm for sorting records into a specified order. Practical examples of sorting abound: a company database may contain records of employees that need to be sorted into alphabetical order of surname, or chronological order of date of employment; an on-line library catalogue may sort records of its holdings alphabetically by author or title, or numerically by call number; and a computer operating system may sort files by filename, date of creation, or the frequency with which the file is accessed.

In each of these examples, the records contain a sorting *key* — a variable whose values can be compared one with another, in order to determine where each record is to appear in the sorted list. We will assume that these keys are numbers, and that the records are to be sorted so that the keys are in increasing order. We will also assume that if two records have the same key, then it does not matter which record appears first in the sorted list. Essentially, then, we have a list of numbers that need to be sorted into increasing order.

Let n be the length of the list of numbers to be sorted. We saw in Chapter 9 that n items can be arranged in $n!$ ways. An algorithm that sorted the list by checking each permutation in turn to find out whether the numbers were

---

[2]The abuse of notation mentioned earlier reappears here. Strictly speaking, it should be 'f is $O(g)$ where $g(n) = n^k$'.

correctly ordered would surely be highly inefficient (we will see later that this is the case), so it is clear that a better method is needed. Many sorting algorithms have been developed, each of which has advantages and disadvantages depending on the situation. Here we will look at a simple insertion algorithm for sorting.

Let x_1, x_2, ..., x_n be the list of numbers in the order in which they are input. The sorting will be carried out *in place*; each step in the algorithm will involve rearranging the numbers in some way, and the algorithm will terminate when the correct order is attained.

Imagine that we are already part-way through sorting the list, so that the first k numbers, x_1, x_2, ..., x_k, have been put into increasing order. If we now insert the next term, x_{k+1}, into its correct position among the numbers already sorted, then the first $k + 1$ numbers will be in their correct order. This process can be repeated until the entire list has been sorted. The starting point for the algorithm is the unsorted list, for which we can take $k = 1$.

Here is an example showing how the process works, using the list 13, 8, 11, 2, 7 as input:

Input:	13, 8, 11, 2, 7
After the first pass:	8, 13, 11, 2, 7
After the second pass:	8, 11, 13, 2, 7
After the third pass:	2, 8, 11, 13, 7
After the fourth pass:	2, 7, 8, 11, 13

In order to write this process as an algorithm, we need to specify how the insertion point is to be found. For the moment, we will do this very simply, by comparing the item to be inserted with each item in the sorted list in turn until we have found the right place. A little later, we will investigate how we can perform the insertion more efficiently.

Here is the algorithm:

1. Input x_1, x_2, ..., x_n
2. **For** $i = 2$ **to** n **do**
    2.1. *insert* $\leftarrow x_i$
    {*insert* is to be inserted in its correct place in the sorted list x_1, x_2, ..., x_{i-1}.}
    2.2. $j \leftarrow 1$
    2.3. **While** $x_j <$ *insert* and $j < i$ **do**
        2.3.1. $j \leftarrow j + 1$
    2.4. **For** $k = 1$ **to** $i - j$ **do**
        2.4.1. $x_{i-k+1} \leftarrow x_{i-k}$
    2.5. $x_j \leftarrow$ *insert*
3. Output x_1, x_2, ..., x_n

The dominant operations here, from the point of view of the time complexity, occur in the inner loops: the **While-do** at Step 2.3 and the **For-do** at Step 2.4. The **While-do** contains two comparisons (in the **While-do** statement itself), an addition and an assignment (in Step 2.3.1), whereas the **For-do** contains only an assignment. Thus it is the **While-do** that determines the time complexity.

Note that if the fourth approximation is being made, then we need not concern ourselves with whether it is the comparison or the addition that is the dominant operation, because this would only change the time complexity by a constant factor. It follows that we can take the time complexity to be just the number of times the loop is executed.

For a fixed value of i, the worst possible case occurs if x_j is always less than *insert*, causing the **While-do** to be executed $i-1$ times. Now, taking account of the outer **For-do** loop, for which i varies from 2 to n, the total number of times the **While-do** is executed is given by the expression:

$$1 + 2 + 3 + \ldots + (n-1)$$

In Example 7.2.1 in Chapter 7, it was proved using induction that $1 + 2 + 3 + \ldots + n = \dfrac{n(n+1)}{2}$. On replacing n by $n-1$ in this result, we obtain the time complexity of the algorithm:

$$\text{Time complexity} = \frac{(n-1)n}{2} = \frac{1}{2}n^2 - \frac{1}{2}n$$

Applying the theorem from the previous section, we conclude that the algorithm has complexity $O(n^2)$.

The process for finding the place to insert the next term among the terms already sorted is rather inefficient, because it makes no use of the fact that these terms are in increasing order. We can improve this part of the algorithm by employing a *binary search* procedure. The method most people would use to look up a word in a dictionary is perhaps not very different from a binary search — open the dictionary in the middle first, determine in which half the word is to be found, then apply the same procedure to the chosen half, repeating the process until the word is found.

In the present example, the binary search procedure begins by comparing *insert* with a number as near as possible to the middle of the sorted list. If *insert* is less than the number, the search can be restricted to the first half of the list; if it is not, then the search can be restricted to the second half of the list. The restricted list is then divided in half again, and the process is repeated until the insertion point is found.

Here is the modified version of the algorithm, incorporating the binary search procedure:

1. Input $x_1, x_2, \ldots, x_n$
2. **For** $i = 2$ **to** n **do**
   - 2.1. *insert* $\leftarrow x_i$
   - 2.2. *lower_limit* $\leftarrow 1$; *upper_limit* $\leftarrow i$

   {*lower_limit* and *upper_limit* are respectively the leftmost and rightmost positions in the list where *insert* could be inserted.}

   - 2.3. **While** *lower_limit* < *upper_limit* **do**
     - 2.3.1. $j \leftarrow$ (*lower_limit* + *upper_limit*) div 2
     - 2.3.2. **If** *insert* < x_j **then**
       - 2.3.2.1. *upper_limit* $\leftarrow j$
       - **else**
       - 2.3.2.2. *lower_limit* $\leftarrow j + 1$

   {*lower_limit* = *upper_limit* after Step 2.3 has finished execution.}

   - 2.4. **For** $k = 1$ **to** $i - lower\_limit$ **do**
     - 2.4.1. $x_{i-k+1} \leftarrow x_{i-k}$
   - 2.5. $x_{lower\_limit} \leftarrow insert$
3. Output $x_1, x_2, \ldots, x_n$

The time complexity of the modified algorithm will depend on how many times the **While-do** is executed, expressed as a function of i. We explore this problem now.

When the **While-do** is entered for the first time, *lower_limit* = 1 and *upper_limit* = i, and the number of positions where *insert* could be inserted is i. After the steps within the **While-do** loop have been executed once, the number of such positions is about half of its former value. (A careful analysis shows that it might not be *exactly* half; if $i = 9$, for example, then j is set to 5 and the number of positions is reduced to either 4 or 5, depending on whether or not *insert* < x_j. If we are making the 'big O' approximation, the fact that it is not exactly half is unimportant.) With each iteration of the loop, the number of positions is reduced by about half, and the loop terminates when the number reaches 1.

If $i = 16 = 2^4$, for example, the successive values for the number of positions are 8, 4, 2, 1, and there will be four iterations of the loop. More generally, if i is any power of 2, say $i = 2^m$, there will be m iterations.

If i is not a power of 2, we can still write $i = 2^m$ and then express m in terms of i, but the result will not be an integer. The number m satisfying the equation $i = 2^m$ is the *base 2 logarithm* of i. The base 2 logarithm of i is sometimes written $\log_2 i$, but base 2 logarithms arise so frequently in the analysis of algorithms that another notation, $\lg i$, is in common use. This leads to the following definition.

Definition

If n is a positive number, then $\lg n$ is the number m that satisfies the equation $n = 2^m$.

For example, $\lg 8 = 3$ (because $2^3 = 8$), and $\lg 16 = 4$ (because $2^4 = 16$). Without the aid of a calculator, we can say that $\lg 10$ must lie somewhere between 3 and 4. If a numerical approximation to $\lg 10$ is needed, it can be obtained from a calculator with a natural logarithm key by using the formula:

$$\lg n = \frac{\ln n}{\ln 2}$$

where $\ln$ denotes the natural logarithm function.

We see now that the number of iterations of the **While-do** in the sorting algorithm incorporating the binary search is approximately $\lg i$.

It is now a straightforward exercise to obtain a 'big O' expression for the time complexity of the algorithm. The **For-do** at Step 2 is iterated $n - 1$ times, with i ranging from 2 to n. We can express the time complexity as:

$$\lg 2 + \lg 3 + \ldots + \lg n$$

For 'big O' purposes, we can use the fact that each term in the above expression is less than or equal to $\lg n$:

$$\lg 2 + \lg 3 + \ldots + \lg n \leq \lg n + \lg n + \ldots + \lg n = n \lg n$$

Therefore the time complexity of the algorithm is $O(n \lg n)$.

How does this algorithm compare with the earlier version, which had time complexity $O(n^2)$? In order to answer this and related questions, we need to know how the growth rate of $\lg n$ compares with the growth rates of the monomial functions n^k. The answer is provided by the following results:

- $\lg n$ is $O(n)$
- n is not $O(\lg n)$

The proof of the first result is left as an exercise. In order to prove that n is not $O(\lg n)$, we need to show that there cannot be a number c such that $n \leq c \lg n$ for all sufficiently large values of n. We can make this very plausible by writing the inequality in the form:

$$\frac{n}{\lg n} \leq c$$

If a number c existed with this property, then $\dfrac{n}{\lg n}$ would not increase without bound as n increased. However, if we evaluate $\dfrac{n}{\lg n}$ for various values of n, we are strongly led to suspect that it does increase without bound. We will outline a proof of this result in the exercises.

These results show that $\lg n$ grows more slowly than n. In fact, it can be proved that $\lg n$ grows more slowly than n^k, where k can be any positive number, no matter how small.

Now that we have established the growth rate of $\lg n$, we can answer the question about the time complexities of the two sorting algorithms. Because $\lg n$ grows more slowly than n, it follows that $n \lg n$ must grow more slowly than n^2. An idea of the relative growth rates of these two functions can be obtained from Table 13.2, where we see that $n \lg n$ is very much smaller than n^2 when n is large. We conclude that the sorting algorithm is faster (in the 'big O' sense) when the insertion is carried out using the binary search procedure.

n	1	2	3	4	5	10	20	50	100	1000	10000	100000
$n \lg n$	0	2	4.75	8	11.6	33.2	86.4	282	664	9966	132877	1.66×10^6
n^2	1	4	9	16	25	100	400	2500	10000	10^6	10^8	10^{10}

Table 13.2

13.6 *Tractable and intractable algorithms*

At this stage, we know how to estimate the time complexity of algorithms and use the results to compare the running times of different algorithms, at least in simple cases where the time complexity is a function whose growth rate we know. We would now like to extend the list of functions that could arise as time complexities, and look at the problem of classifying algorithms according to whether they can be executed within a realistic period of time.

A good way to start is to look at some of the algorithms we have studied in earlier chapters. The algorithm we encountered in Chapter 10 for finding an Eulerian circuit in an Eulerian graph is one such algorithm. Here it is again:

1. Input an Eulerian graph G, with vertex set $V(G) = \{v_1, v_2, \ldots, v_m\}$ and edge set $E(G) = \{e_1, e_2, \ldots, e_n\}$.
2. *circuit* ← v_1; *unused_edges* ← $E(G)$
3. **While** *unused_edges* ≠ ∅ **do**
   - 3.1. *insertion_point* ← first vertex in *circuit* with unused edges incident to it
   - 3.2. v ← *insertion_point*; *new_circuit* ← v
   - 3.3. **Repeat**
      - 3.3.1. e ← first element of *unused_edges* incident to v
      - 3.3.2. v ← vertex adjacent to v via edge e
      - 3.3.3. *new_circuit* ← *new_circuit*, e, v
      - 3.3.4. *unused_edges* ← *unused_edges* − $\{e\}$

      until no element of *unused_edges* is incident to v
   - 3.4. *circuit* ← (*circuit* before *insertion_point*), *new_circuit*, (*circuit* after *insertion_point*)
4. Output *circuit*

The steps within the **While-do** at Step 3 are executed n times, where n is the number of edges in the graph. With each iteration of the **While-do**, the steps within the **Repeat-until** at Step 3.3 are executed m times in the worst case, where m is the number of vertices in the graph. (We could be more precise here, and allow for the fact that the number of unused edges decreases by 1 with each iteration of the **While-do**, but this would make no difference to the final 'big O' answer.) The statements within the **Repeat-until** will therefore be executed $O(mn)$ times altogether.

Now, looking carefully at the steps within the **Repeat-until**, we see that Step 3.3.1 involves a search for the first element of *unused_edges* that is incident to v. In the worst case, this would mean examining all of the elements of *unused_edges* until the required edge is found, which in the worst case would take n comparisons. The time complexity of the algorithm is therefore $O(mn^2)$.

How do we measure the size of the input? A careful answer to this question would mean examining the way in which the graph is to be input (as an adjacency matrix, for example), and relating the number of edges and vertices of the graph to the number of bits in the input. In the present example, however, we will disregard this complication and express the complexity as a function of n, the number of edges.

Note that n cannot be less than $m - 1$ in a connected graph, because $n = m - 1$ for a tree, and removing an edge from a tree produces a disconnected graph. Therefore either m and n have the same order of magnitude, or m has a smaller order of magnitude than n. Hence we can write the time complexity of the Eulerian circuit algorithm as $O(n^3)$.

We now turn our attention to the corresponding problem for Hamiltonian circuits. Recall that a Hamiltonian circuit is a circuit that visits each vertex exactly once and then returns to the starting vertex. When we studied Hamiltonian circuits in Chapter 10, we did not provide an algorithm for finding a Hamiltonian circuit in a Hamiltonian graph. One way of searching for a Hamiltonian circuit in a graph would be to test every permutation of the vertices to find out whether the permutation corresponds to a circuit. This doesn't seem to be a very efficient way of solving the problem, but just how inefficient is it?

We know that there are $n!$ permutations of the n vertices. Even without analysing how long it would take to check a particular permutation, the process of generating all of the permutations will be very time consuming if n is large. We can prove this by establishing the following results:

- 2^n is not $O(n^k)$, for any value of k
- 2^n is $O(n!)$

We have already established that n is not $O(\lg n)$. This means that there is no number k such that $n \le k \lg n$ for all sufficiently large values of n. We can obtain an inequality that is equivalent to $n \le k \lg n$ by raising 2 to the power of each side of the inequality:

$$2^n \le 2^{k \lg n}$$

Now:

$$2^{k \lg n} = (2^{\lg n})^k = n^k$$

and so the original inequality is equivalent to the following one:

$$2^n \le n^k$$

Therefore we can say that there is no number k such that $2^n \le n^k$ for all sufficiently large values of n. This proves the first result.

To prove the second result, we can use the definition of $n!$ to obtain the following inequality:

$$n! = n(n-1)(n-2)\ldots \times 2 \times 1 \ge 2 \times 2 \times 2 \times \ldots \times 2 \times 1 = 2^{n-1}$$

Therefore $2^n \le 2n!$, and so 2^n is $O(n!)$ (taking $c = 2$ in the definition of 'big O').

What these results tell us is that an algorithm that has time complexity $O(2^n)$ or $O(n!)$ will always be slower than one that has a polynomial for its time complexity.

In fact, it is possible to say rather more. From a practical point of view, an algorithm with a polynomial time complexity and 'realistic' input can usually be executed within a reasonable amount of time if a sufficiently powerful machine is available. Conversely, if the time complexity is $O(2^n)$ or $O(n!)$, then the execution time grows so rapidly as a function of the size of the input that it will usually be impracticable to run the algorithm when once the input exceeds a very modest size. In this context, the following definition is useful.

Definition

An algorithm is *tractable* if its time complexity is $O(n^k)$ for some number k, where n is the size of the input (in number of bits).

Algorithms with polynomial time complexities are tractable. Algorithms with exponential time complexities such as 2^n are intractable, as are algorithms with time complexity $n!$. Thus our algorithm for finding an Eulerian circuit is tractable, while the search for a Hamiltonian circuit by checking each permutation of the vertices is intractable.

There are many problems of practical interest for which no tractable algorithm is known. The Hamiltonian circuit problem is one example; while it is true that there are more efficient ways of going about solving the problem than the way we proposed, no-one has yet constructed an algorithm that performs the task in polynomial time. Other problems for which there is no known tractable algorithm are the Travelling sales representative problem (which we met in Chapter 11), the factorisation of a natural number into primes, the problem of determining whether two graphs are isomorphic, and the problem of determining whether a logical proposition is a tautology (or whether it is a contradiction). In connection with the last of these, notice that constructing a truth table is an intractable algorithm, because a truth table for an expression with n variables has 2^n rows. If a logical expression has a large number of variables, constructing a truth table for it becomes infeasible.

It might seem that we are being rather dogmatic in claiming that algorithms that cannot run in polynomial time will also take an impossibly long time to run on a real computer. After all, computer technology is advancing rapidly, and it could be argued that computers in the future might be able to execute algorithms that seem impractical at present.

A look at some figures should persuade you that this argument cannot be sustained. Suppose we have at our disposal a '1 gigaflop' machine that performs 10^9 floating-point operations (operations with real numbers) per second. If we take a floating-point operation to be the 'time-consuming operation' in our analysis of time complexity, we can calculate how long algorithms with various time complexities would take to run as a function of the input size. The results are shown in Table 13.3.

Time complexity	Input size		
	10	100	1000
n	10^{-8} s	10^{-7} s	10^{-6} s
$n \lg n$	3×10^{-8} s	7×10^{-7} s	10^{-5} s
n^2	10^{-7} s	10^{-5} s	10^{-3} s
n^3	10^{-6} s	10^{-3} s	1 s
n^4	10^{-5} s	0.1 s	17 min
n^5	10^{-4} s	10 s	278 days
n^{10}	10 s	3×10^3 y	3×10^{13} y
2^n	10^{-6} s	4×10^{13} y	10^{284} y
$n!$	4×10^{-3} s	10^{141} y	10^{2551} y

Table 13.3

Even a computer of the future that could run billions of times faster would scarcely make a dent in the astronomical figures in the lower right-hand corner of Table 13.3. In the face of figures like these, the fact that we have made a few approximations along the way in order to obtain them scarcely seems relevant.

In fact, there are a few exceptions to the situation as we have portrayed it here. One exception arises in the branch of mathematics known as linear programming, which deals with certain types of problems in which a quantity (such as profit) is to be maximised subject to certain constraints (such as limitations on the availability of labour and materials). One algorithm in widespread use for solving linear programming problems (the 'simplex' method) is actually known to be intractable in the sense we have defined it. In this case, it is our second approximation that is to blame; the linear programming problems that arise in practice can always be solved much more efficiently than the theoretical 'worst case'.

Isolated exceptions such as this, however, do not substantially alter the general conclusion, which is that an intractable algorithm will almost always be impractical to implement on a computer except when the input is small. If a problem of practical importance has no known tractable algorithm, we must make a choice. In some problems, an approximate solution may be acceptable; for example, in the Travelling sales representative problem, we might design an algorithm to find a circuit that is fairly short but not necessarily the shortest possible. Otherwise, we must accept that we have reached the limits of current knowledge, and that the problem cannot be solved until such time as a tractable algorithm has been developed. It is with these sobering thoughts that we conclude our study of algorithms.

Exercises

1. Identify the dominant operation in the algorithm in Chapter 1, Exercise 10. Hence obtain an expression for the time complexity of the algorithm as a function of n.

2. For the algorithm you obtained as the answer to Chapter 1, Exercise 6, identify the dominant operation, and hence obtain an expression for the time complexity in the worst case.

3. Trace the algorithm in Section 13.3 for evaluating x^n, with $x = 2$ and $n = 1011_2$.

4. Find the number of multiplications needed when x^{15} is calculated as:
   (a) $((x^2)^2)^2(x^2)^2x^2x$
   (b) $((x^2)^2x)^3$

5. Two algorithms, A and B, have time complexities f and g respectively, where $f(n) = 3n^2 + n - 4$ and $g(n) = n^2 + 2n + 2$. Without using the 'big O' approximation, determine which algorithm is faster.

6. Use the definition of 'big O' to show that f is $O(g)$ for the following functions f and g:

(a) $f(n) = 8n$, $g(n) = n$
(b) $f(n) = 4n - 1$, $g(n) = n$
(c) $f(n) = 2n^2 + 5n - 3$, $g(n) = n^2$
(d) $f(n) = n^2$, $g(n) = n^3$

7. Let $f(n) = n^3$ and $g(n) = n^2$. Use the definition of 'big O' to show that f is not $O(g)$.

8. Using any of the results obtained in the text, rank the following complexities in increasing order in the sense of 'big O'. Identify any complexities that are equivalent in the 'big O' sense.

$$n^3 + 4n, \ n^2 \lg n, \ 3 \times 2^n, \ 10n + 13, \ n^2 + n \lg n, \ n - 4, \ n \sqrt{n}$$

9. Trace the insertion sort algorithm with the binary search procedure, using the sequence 8, 5, 7 as input.

10. Prove that $\lg n$ is $O(n)$. (Use the fact that $2^n > n$ for $n \geq 0$, which you proved in Chapter 7, Exercise 20(b).)

11. Prove that $\dfrac{n}{\lg n}$ increases without bound, using the following steps:

1. In the result $2^n > n$, which we already know to be true when $n \geq 0$, replace n by $k - 1$, and hence show that $2^k - k > k - 2$ when $k \geq 1$.

2. By substituting $n = 2^{2^k}$ in $\dfrac{n}{\lg n}$, show that $\dfrac{n}{\lg n} > 2^{k-2}$.

3. Deduce from Step 2 that $\dfrac{n}{\lg n}$ must increase without bound as n increases.

12. Consider the problem of obtaining a 'big O' estimate of the time complexity of Prim's algorithm in Chapter 11 as a function of the number of vertices, n, and the number of edges, m, in the graph.

(a) Obtain a 'big O' estimate of the time complexity of Step 3.1 of Prim's algorithm, assuming that the process of determining whether an edge is incident to exactly one vertex could require that each vertex be investigated.

(b) Hence obtain a 'big O' estimate of the time complexity of Prim's algorithm.

13. Comment on the tractability of an algorithm that lists all of the subsets of a set with n elements.

Answers to Exercises

CHAPTER 1

1.

   1. Input the number of values n

   2. Input the list of numbers $x_1, x_2, \ldots, x_n$

   3. $min \leftarrow x_1$

   4. $position \leftarrow 1$

   5. **For** $i = 2$ **to** n **do**

      5.1. **If** $x_i < min$ **then**

         5.1.1. $min \leftarrow x_i$

         5.1.2. $position \leftarrow i$

   6. Output $min, position$

2.

   1. Input $hours, minutes, seconds$

   2. $answer \leftarrow 3600 \times hours + 60 \times minutes + seconds$

   3. Output $answer$

3.

   1. Input n

   2. $sum \leftarrow 0$

   3. **For** $i = 1$ **to** n **do**

      3.1. $sum \leftarrow sum + i^2$

   4. Output sum

4.

   1. Input $price, tendered$

   2. **If** $tendered < price$ **then**

      2.1. Output 'Amount tendered is not sufficient.'

      else

      2.2. $change \leftarrow tendered - price$

      2.3. Output $change$

5.

   1. Input $income$

   2. **If** $income \leq 5400$ **then**

      2.1. $tax \leftarrow 0$

      else if $income \leq 20700$ **then**

      2.2. $tax \leftarrow 0.2 \times (income - 5400)$

      else if $income \leq 36000$ **then**

      2.3. $tax \leftarrow 3060 + 0.38 \times (income - 20700)$

      else if $income \leq 50000$ **then**

      2.4. $tax \leftarrow 8874 + 0.46 \times (income - 36000)$

      else

      2.5. $tax \leftarrow 15314 + 0.47 \times (income - 50000)$

   3. Output tax

6.

1. Input $x_1, x_2, \ldots, x_n$
2. $i \leftarrow 1$
3. $order\_ok \leftarrow$ true
4. **While** $i < n$ and $order\_ok$ **do**
      4.1. **If** $x_i > x_{i+1}$ **then**
          4.1.1. $order\_ok \leftarrow$ false
      4.2. $i \leftarrow i + 1$
5. **If** $order\_ok$ **then**
      5.1. Output 'Numbers are in order.'
      else
      5.2. Output 'Numbers are out of order.'

7.

1. Input x, n
2. $answer \leftarrow 1$
3. **For** $i = 1$ **to** n **do**
      3.1. $answer \leftarrow answer \times x$
4. Output $answer$

8. (a)

Step	n	d	Output
1	8678	–	–
2	8678	4	–
3	8678	4	–
3.1	29	4	–
3.2	29	2	–
3	29	2	–
3.1	11	2	–
3.2	11	2	–
3	11	2	–
3.1	2	2	–
3.2	2	1	–
3	2	1	–
4	2	1	2

(b) 1, 2, 3, 4, 5, 6, 7, 8, 9

9. (a) 2   **(b)** 0   **(c)** The process never terminates.   **(d)** No.

10. (a)

Step	n	$answer$	Output
1	4	–	–
2	4	4	–
3	4	4	–
3.1	3	4	–
3.2	3	12	–
3	3	12	–
3.1	2	12	–
3.2	2	24	–
3	2	24	–
3.1	1	24	–
3.2	1	24	–
3	1	24	–
4	1	24	24

(b) Yes. The steps are well defined. The execution sequence is well defined. Step 3 is executed $n - 1$ times, so the process always terminates.

11.
  1. Input n
  2. Input $c_1 c_2 ... c_n$
  3. $excess\_left \leftarrow 0$; $i \leftarrow 1$
  4. **While** $i \leq n$ and $excess\_left \geq 0$ **do**
      4.1. **If** $c_i = $ '(' **then**
              4.1.1. $excess\_left \leftarrow excess\_left + 1$
              else if $c_i = $ ')' **then**
              4.1.2. $excess\_left \leftarrow excess\_left - 1$
      4.2. $i \leftarrow i + 1$
  5. **If** $excess\_left = 0$ **then**
      5.1. Output 'Parentheses paired correctly.'
      else
      5.2. Output 'Parentheses paired incorrectly.'

12. (a) 1st pass: 1111111111
        2nd pass: 1010101010
        3rd pass: 1000111000
        4th pass: 1001111100
        5th pass: 1001011101
        6th pass: 1001001101
        7th pass: 1001000101
        8th pass: 1001000001
        9th pass: 1001000011
        10th pass: 1001000010
    (b) $a_i = 1$ if i is a perfect square, $a_i = 0$ otherwise.

13. (a) 7, 22, 11, 34, 17, 52, 26, 13, 40, 20, 10, 5, 16, 8, 4, 2, 1

CHAPTER 2

1. $3 \times 10^2 + 9 \times 10^1 + 4 \times 10^0 + 2 \times 10^{-1} + 7 \times 10^{-2}$

2. (a) 101_{10} (b) 87.6875_{10}

3. (a) 1100111010_2 (b) 0.01011_2 (c) 11001000100.0011_2
   (d) -111010111.01_2

4. (a) 0.00110_2 (b) 1101.01111_2

5. (a) The integer is multiplied by 2.
   (b) The integer is multiplied by 2, and 1 is added.

6. (a) 2509_{10} (b) 387.328125_{10} (c) 3182_{10} (d) 762.5_{10}

7. (a) 3726_{10} (b) 182095_{10} (c) 36018_{10} (d) 14979295_{10}

8.

1. Input b {b is the base}
2. Input n {n is in base b, and is the number to be converted}
3. $d \leftarrow$ number of digits in n
4. *answer* $\leftarrow$ 1st digit of n {counting from the left}
5. **For** $i = 2$ **to** d **do**
    5.1. $k \leftarrow$ ith digit of n
    5.2. *answer* $\leftarrow$ *answer* $\times b + k$
6. Output *answer*

9. (a) 7402_8 **(b)** 443.74_8

10. (a) $746B_{16}$ **(b)** $1B32.94_{16}$

11. (a) 1646_8, $3A6_{16}$ **(b)** 305.16_8, $C5.38_{16}$

12. (a) 10100111_2 **(b)** 11001.110011_2 **(c)** 100100111011_2
    (d) 10101101.000111_2

13. (a) 74, 1D   **(b)** CE, 73

14. (a) 11001011_2 **(b)** 1111000_2 **(c)** 1000110_2 **(d)** 1011_2
    (e) 11110111_2 **(f)** 1000100010_2 **(g)** 110_2 **(h)** 111.010_2

CHAPTER 3

1. (a) 00101100_2 **(b)** 10010111_2

2.

1. Input $b_1 b_2 \ldots b_n$
2. $i \leftarrow n$
3. **While** $b_i = 0$ **do**
    3.1. $c_i \leftarrow 0$
    3.2. $i \leftarrow i - 1$
4. $c_i \leftarrow 1$
5. **For** $j = 1$ **to** $i - 1$ **do**
    5.1. $c_j \leftarrow 1 - b_j$
6. Output $c_1 c_2 \ldots c_n$

4. (b) The 10's complement of b is $10^k - b$, where k is the number of digits in b. Therefore a plus the 10's complement of b equals $a + 10^k - b$. Ignoring the leftmost 1 is equivalent to subtracting 10^k, giving an answer of $a - b$.

5. (a) 01110100 01101011   **(b)** 11100000 00100101

6. 48

9. $0.11011101001001 \times 2^{10}$; 01000100 11101110 10010010 00000000

10. (a) 01000110 01011100 00011001 10000001
    (b) 10111101 11001111 01000001 11110010

11. 01000000 01001110 11101001 00100000

12. 0.56×10^{-308} to 0.18×10^{309}

13. Because the representation of real numbers is not exact, the condition $x = 10.0$ may never be satisfied.   Replace $x = 10.0$ with $|x - 10.0| < 0.05$.

14. (a) 0.7097×10^8 **(b)** 0.6925×10^3 **(c)** 0.6700×10^{-2}

15. If the two numbers have equal exponents and almost equal mantissas, so that the difference between the mantissas is less than 0.1, then the normalised mantissa of the answer will have one or more trailing zeros that are not significant. (An example where this occurs is Exercise 14(c).)

CHAPTER 4

1. (a) $(p \wedge \neg q) \vee q$ **or**
  (b) $\neg(p \to q)$ **not**
  (c) $p \leftrightarrow (\neg q \vee r)$ **if and only if**
  (d) $[(p \wedge \neg q) \wedge r] \to [\neg(q \vee \neg r) \to (p \vee r)]$ first **implies**

2. (a) It is not snowing and I will go skiing.
  (b) If it is snowing then I will go skiing.
  (c) If I will not go skiing then it is snowing.
  (d) It is snowing or I will not go skiing, and it is snowing.

3. (a)

p	q	$p \oplus q$
T	T	F
T	F	T
F	T	T
F	F	F

(b)

p	q	$p \vee q$	$p \wedge q$	$\neg(p \wedge q)$	$(p \vee q) \wedge \neg(p \wedge q)$
T	T	T	T	F	F
T	F	T	F	T	T
F	T	T	F	T	T
F	F	F	F	T	F

4. (a) If an error message is not generated, then the input file exists.
  If an error message is generated, then the input file does not exist.
  (b) If my program cannot run, then a graphics driver is not available.
  If my program can run, then a graphics driver is available.
  (c) If my program produces correct output, then it contains no bugs.
  If my program doesn't produce correct output, then it contains bugs.

5. If I will go skiing then it is snowing.
  If I will not go skiing then it is not snowing.

6. (a)

p	q	$\neg(p \vee \neg q) \vee p$
T	T	T
T	F	T
F	T	T
F	F	F

Neither.

(b)

p	q	$[p \rightarrow (p \wedge q)] \rightarrow \neg q$
T	T	F
T	F	T
F	T	F
F	F	T

Neither.

(c)

p	q	$(p \wedge q) \leftrightarrow (\neg p \vee \neg q)$
T	T	F
T	F	F
F	T	F
F	F	F

Contradiction.

(d)

p	q	r	$[(p \wedge r) \vee (q \wedge r)] \rightarrow (p \rightarrow \neg q)$
T	T	T	F
T	T	F	T
T	F	T	T
T	F	F	T
F	T	T	T
F	T	F	T
F	F	T	T
F	F	F	T

Neither.

7. (a) If for a particular set of truth values of the variables P is true, then $P \vee Q$ must be true, so there is no need to find the truth value of Q.

(b) (i)

p	q	r	$[\neg(p \wedge q) \wedge (p \vee \neg r)] \wedge [(p \wedge r) \vee \neg q]$
T	T	T	F
T	T	F	F
T	F	T	T
T	F	F	T
F	T	T	F
F	T	F	F
F	F	T	F
F	F	F	T

(ii)

p	q	r	$\neg[\neg p \wedge (q \vee r)] \vee (\neg p \wedge \neg r)$
T	T	T	T
T	T	F	T
T	F	T	T
T	F	F	T
F	T	T	F
F	T	F	T
F	F	T	F
F	F	F	T

10. (a) p **(b)** $p \wedge q$ **(c)** $\neg p \wedge \neg q$ **(d)** $\neg p \vee \neg q \vee \neg r$

11. If $x < 3$ or $x \geq 6$ **then** …

12.

1. $n \leftarrow 0$
2. $term \leftarrow 1$
3. $sum \leftarrow 0$
4. $n \leftarrow n + 1$
5. $term \leftarrow term / 2$
6. $sum \leftarrow sum + term$
7. **While** $term \geq 0.001$ and $n \neq 100$ **do**
   7.1. $n \leftarrow n + 1$
   7.2. $term \leftarrow term / 2$
   7.3. $sum \leftarrow sum + term$

13. (a) Tautology. **(b)** Contradiction. **(c)** Tautology.

14. $[(p \rightarrow q) \wedge q] \rightarrow p$; not valid.

15. $[(p \vee \neg q) \wedge (q \rightarrow \neg p)] \rightarrow p$; not valid.

16. $\neg(\neg p \wedge \neg q)$

17. (a) $p \mid p$ **(b)** $(p \mid q) \mid (p \mid q)$ **(c)** $(p \mid p) \mid (q \mid q)$

18. (a) $\exists x \, P(x)$, where $P(x)$ denotes '$x^2 - 3x + 2 = 0$'; (or, more simply,
   $\exists x \, (x^2 - 3x + 2 = 0)$). True.
   (b) $\forall x \, \exists y \, P(x, y)$, where $P(x, y)$ denotes '$x = y^2$'. False.

19. (a) $\forall x \, (x^2 - 3x + 2 \neq 0)$. For every number x, $x^2 - 3x + 2 \neq 0$.
   (b) $\exists x \, \forall y \, (x \neq y^2)$. There is a number x such that, for every number y,
   $x \neq y^2$.

20. (a) $\exists b \, B(p, b)$ **(b)** $\exists p \, B(p, b)$ **(c)** $\forall p \, [\neg B(p, b)]$
   (d) $\exists b \, \exists c \, [(b \neq c) \wedge B(p, b) \wedge B(p, c)]$
   (e) $\forall p \, \forall q \, \forall b \, \{[B(p, b) \wedge B(q, b)] \rightarrow (p = q)\}$
   (f) $\forall b \, [\neg O(b)]$ **(g)** $\forall b \, \{O(b) \rightarrow [\exists p \, B(p, b)]\}$
   (h) $\exists b \, [B(p, b) \wedge O(b)]$

21. (a) True. **(b)** 'This statement doesn't have five words.' True.

23. Ask the second person: 'Are you using the software on the network?'
   Ask the third person: 'Are you a student?'

CHAPTER 5

1. (a) {a, e, i, o, u}   **(b)** {12, 15, 18}   **(c)** {1, 6, 11, 16, 21, ...}

2. (a) $\{x \in \mathbf{N}: x \le 20 \text{ and } x \text{ is divisible by } 4\}$
   (b) {x: x is a string of 3 bits}
   (c) $\{x: x = y^2 \text{ for some natural number } y\}$

3. (a), (c), (d), (e), (g) and (j) are true, the others are false.

4. (a) {9, 11}   **(b)** {3, 6, 8, 9, 10, 11, 12}   **(c)** {2, 4, 6, 8, 10, 12}
   (d) {3, 6, 9}   **(e)** {1, 2, 4, 5, 7, 8, 10, 11}

8. 20

9. $|A \cup B \cup C| = |A| + |B| + |C| - |A \cap B| - |A \cap C| - |B \cap C| + |A \cap B \cap C|$

10. Yes: $\varnothing$ has exactly one subset, namely $\varnothing$ itself.

11. (a) {(a, p), (a, q), (b, p), (b, q), (c, p), (c, q)}
   (b) {(a, a), (a, b), (a, c), (b, a), (b, b), (b, c), (c, a), (c, b), (c, c)}
   (c) {(p, p, p), (p, p, q), (p, q, p), (p, q, q), (q, p, p), (q, p, q), (q, q, p), (q, q, q)}

12. (a) {entrées} × {main courses} × {desserts}
   (b) $L^3 \times D^3$, where L is the set of letters and D is the set of digits.
   (c) $\{\text{heads, tails}\}^3$

13. (a) 0010 1101 0001 0010   **(b)** {0, 2, 5, 6, 8, 9, 10, 12, 15}
   (c) 0010 0000 0000 0101
       1011 1101 0111 1111
       1100 1011 1001 0010
       0101 0110 1110 1000

14.

1. Input n
2. Input $a_1a_2...a_n$ {string of bits representing A}
3. Input $b_1b_2...b_n$ {string of bits representing B}
4. **For** $i = 1$ **to** n **do**
   4.1. **If** $a_i = 1$ and $b_i = 0$ **then**
       4.1.1. $c_i \leftarrow 1$
       else
       4.1.2. $c_i \leftarrow 0$
5. Output $c_1c_2...c_n$

15. (a)

	1	2	3	4	5
1	F	F	T	T	F
2	T	T	F	T	F
3	F	F	F	F	T
4	F	F	F	F	F
5	F	T	F	F	T

(b)

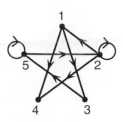

16. (a)

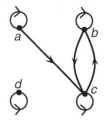

(b) Reflexive: the matrix has T along its principal diagonal. Not symmetric: aRc, but c is not related to a. Not transitive: aRc and cRb, but a is not related to b.

17. By changing the order of the elements in the underlying set, different matrices representing the same relation can be produced.

18. (a) Irreflexive, symmetric.   (b) Irreflexive, antisymmetric.
    (c) Irreflexive, antisymmetric, transitive.   (d) Reflexive, symmetric, transitive.   (e) Reflexive, symmetric, transitive.   (f) Reflexive, antisymmetric, transitive.

19. (d) and (e) are equivalence relations. Equivalence classes for (d): all sets of the form $\{x, -x\}$ where x is a positive real number, and the set $\{0\}$. Equivalence classes for (e): all sets of the form $\{x: n \leq x < n + 1\}$ where $n \in \mathbf{N}$; all sets of the form $\{x: -n < x \leq -n + 1\}$; and the set $\{x: -1 < x < 1\}$.

20. Only (f) is a partial order relation.

21. (b) M_3

CHAPTER 6

1. (a) Domain = $\mathbf{R}$, codomain = $\mathbf{R}$, range = $\mathbf{R}$ (b) Not well defined.
   (c) Not well defined.   (d) Domain = $\mathbf{N}$, codomain = $\mathbf{N}$, range = $\mathbf{N}$
   (e) Domain = $\mathbf{R} - \{0\}$, codomain = $\mathbf{R}$, range = $\mathbf{R} - \{0\}$
   (f) Not well defined.   (g) Domain = $\mathbf{N}$, codomain = $\mathbf{J}$,
   range = $\{0, 1, 2, 3, 4, 5\}$ (h) Not well defined.

2. (a) One-to-one, onto.   (b) Onto.   (c) One-to-one.   (d) Onto.
   (e) One-to-one.   (f) One-to-one only if $|A| \leq 1$.

3.

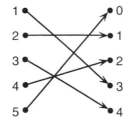

One-to-one and onto.

4. (b) Not one-to-one — two different strings can have the same check digit. Onto — every digit can occur as a check digit.

   (c) No; f is not one-to-one, so an incorrect code could have the same check digit as the correct code.

5. (a) $(f \circ f)(x) = 4(4x - 3) - 3$ **(b)** $(f \circ g)(x) = 4(x^2 + 1) - 3$

   (c) $(g \circ f)(x) = (4x - 3)^2 + 1$ **(d)** $(f \circ h)(x) = \begin{cases} 1 & \text{if } x \geq 0 \\ -3 & \text{if } x < 0 \end{cases}$

   (e) $(h \circ f)(x) = \begin{cases} 1 & \text{if } x \geq \frac{3}{4} \\ 0 & \text{if } x < \frac{3}{4} \end{cases}$ **(f)** $(g \circ h)(x) = \begin{cases} 2 & \text{if } x \geq 0 \\ 1 & \text{if } x < 0 \end{cases}$

   (g) $(h \circ g)(x) = 1$

6. (a) $f^{-1}: \mathbf{R} \to \mathbf{R}, \ f^{-1}(x) = \dfrac{x - 2}{3}$

   (b) No inverse — not one-to-one (and not onto).

   (c) g is its own inverse.

   (d) $h^{-1}: S \to S, \ h^{-1}(s) =$ the string obtained by moving the first character to the end of the string.

7. (a) $g \circ f: X \to \mathbf{N}, \ (g \circ f)(x) =$ age of student with name x

     $f^{-1}: Y \to X, \ f^{-1}(y) =$ name of student with ID number y

   (b) g is not onto, since there are elements of $\mathbf{N}$ that cannot possibly correspond to ages of students. (g is presumably not one-to-one either, since we would expect to be able to find two different students of the same age.)

8. (a) 'BARBARA HILL', 'barbara hill'

   (b) No, since $lwr \circ upr$ is not the identity function.

9. (a) (i) Yes; two different numbers cannot correspond to the same English word.

     (ii) No; some character strings are not words for numbers from 1 to 20.

     (iii) No, since f is not onto.

   (b) (i) 5   **(ii)** Does not exist.   **(iii)** Does not exist.

     (iv) 'five'.

10. (a) Not one-to-one — two different natural numbers can have the same digital root, e.g. 2 and 11 both have a digital root of 2. Not onto — the range of f is $\{1, 2, 3, 4, 5, 6, 7, 8, 9\}$, which is a proper subset of the codomain.

   (b) No, since f is not one-to-one (and not onto).   **(c)** $f \circ f = f$

CHAPTER 7

1. (a) 7, 11, 15   **(b)** 3, 6, 10   **(c)** 1, 3, 7   **(d)** 6, 15, 31

2. 3, 5, 11

3. (a)

     1. Input m
     2. $t \leftarrow 3$
     3. Output t
     4. **For** $n = 2$ **to** m **do**
         4.1. $t \leftarrow t + 4$
         4.2. Output t

  (b)

     1. Input m
     2. $t \leftarrow 1$
     3. Output t
     4. **For** $n = 2$ **to** m **do**
         4.1. $t \leftarrow t + n$
         4.2. Output t

4. $2^0 = 1$; $2^n = 2 \times 2^{n-1}$ $(n > 0)$

5. $t(1) = 1$; $t(n) = t(n-1) + 2n - 1$ $(n > 1)$

6. Recursive: $t(1) = 2$; $t(n) = t(n-1) + 5$ $(n > 1)$.
Non-recursive: $t(n) = 5n - 3$

7. $t(1) = 1$; $t(n) = 2t(n-1) + 3$ $(n > 1)$

9. (a) 25   **(b)** 990   **(c)** 120   **(d)** $n(n-1)$

10.

Step	n	f	i	Output
1	4	–	–	–
2	4	1	–	–
3	4	1	1	–
3.1	4	1	1	–
3	4	1	2	–
3.1	4	2	2	–
3	4	2	3	–
3.1	4	6	3	–
3	4	6	4	–
3.1	4	24	4	–
4	4	24	4	24

13. (a) 22   **(b)** 175   **(c)** 220

17. (a) $t(n) = 5n - 2$ **(b)** $t(n) = \dfrac{1}{n}$ **(c)** $t(n) = n^2 + 1$

18. All natural numbers greater than or equal to 5.

19. All odd natural numbers.

21. (a) $value(1) = 2$; $value(i) = value(i-1) + i + 1$ $(i > 1)$
(b) $value(1) = 1$; $value(i) = value(i-1) + (2i-1)^2$ $(i > 1)$

22. (a) Function $t(n)$:
    1. **If** $n = 1$ **then**
        1.1. $t \leftarrow 3$
    else
        1.2. $t \leftarrow t(n-1) + 4$
(b) Function $t(n)$:
    1. **If** $n = 1$ **then**
        1.1. $t \leftarrow 1$
    else
        1.2. $t \leftarrow t(n-1) + n$

23. (c) Algorithm *write_sequence(m, last_term)*:
    1. **If** $m = 1$ **then**
        1.1. *last_term* $\leftarrow 0$
    else
        1.2. *write_sequence(m − 1, secondlast_term)*
        1.3. *last_term* $\leftarrow 2$ *secondlast_term* $+ 1$
    2. Output *last_term*
(d) Algorithm *write_sequence(m, last_term)*:
    1. **If** $m = 1$ **then**
        1.1. *last_term* $\leftarrow 2$
    else
        1.2. *write_sequence(m − 1, secondlast_term)*
        1.3. *last_term* $\leftarrow$ *secondlast_term* $+ m^2$
    2. Output *last_term*

24. (c)

Step		Copy 1			Copy 2			Copy 3		Output
	m	*last*	*secondlast*	*m*	*last*	*secondlast*	*m*	*last*	*secondlast*	
Initially	3	–	–							
1.2 (hold)										
Initially				2	–	–				
1.2 (hold)										
Initially							1	–	–	
1.1							1	0	–	
2							1	0	–	0
1.2				2	–	0				
1.3				2	1	0				
2				2	1	0				1
1.2	3	–	1							
1.3	3	3	1							
2	3	3	1							3

(d)

Step	Copy 1			Copy 2			Copy 3			
	m	*last*	*secondlast*	m	*last*	*secondlast*	m	*last*	*secondlast*	Output
Initially	3	–	–							
1.2 (hold)										
Initially				2	–	–				
1.2 (hold)										
Initially							1	–	–	
1.1							1	2	–	
2							1	2	–	2
1.2				2	–	2				
1.3				2	6	2				
2				2	6	2				6
1.2	3	–	6							
1.3	3	15	6							
2	3	15	6							15

25. Algorithm *test_for_all_zeros(n, all_zeros)*:
   1. Input b_n {the nth bit}
   2. **If** $n = 1$ **then**
      2.1. **If** $b_1 = 0$ **then**
            2.1.1. *all_zeros* ← true
            else
            2.1.2. *all_zeros* ← false
      else
      2.2. *test_for_all_zeros(n – 1, all_zeros)*
      2.3. **If** *all_zeros* = true and $b_n = 0$ **then**
            2.3.1. *all_zeros* ← true
            else
            2.3.2. *all_zeros* ← false

26. Algorithm *find_number_of_a(n, number_of_a)*:
   1. Input c_n
   2. **If** $n = 1$ **then**
      2.1. **If** $c_1 = $ 'a' **then**
            2.1.1. *number_of_a* ← 1
            else
            2.1.2. *number_of_a* ← 0
      else
      2.2. *find_number_of_a(n – 1, number_of_a)*
      2.3. **If** $c_n = $ 'a' **then**
            2.3.1. *number_of_a* ← *number_of_a* + 1

CHAPTER 8

1. Division by 0 is not defined. Division is a binary operation on **R** – {0}, because the quotient of two non-zero real numbers is always defined, and is a non-zero real number.

2. (a), (c) and (d) are binary operations.

9. (a) Yes.   (b) No.   (c) No.

10. (a) y (b) $x \times y$ (c) $(x + y) \times z$ (d) $(x' + y') + z$

11. (a) $q \wedge [p \vee (\neg p \wedge q)] \equiv q$ (b) $B \cap [A \cup (\overline{A} \cap B)] = B$

12. (a) $x + y$

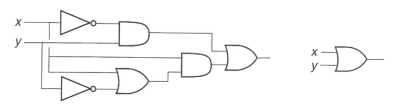

(b) $x'y$

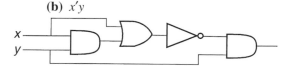 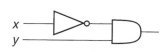

(c) $x' + y' + z'$

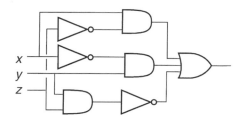 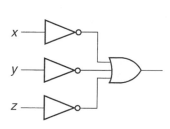

13. (a) xy (b) $x(y + z')$

 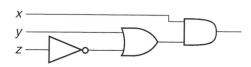

14. $x'y'z + x'yz + xy'z' + xy'z$

15. $x'y'z + x'y'z' + x'yz' + xy'z$

16. (a) $x' + y'$ (b) $x' + y'z'$ (c) $xy + x'z'$ (d) $xy + xz'$
    (e) $x'y + x'w + zw' + y'z'w$
    (f) $xyw' + xzw' + x'yz'w + x'y'zw + x'y'z'w'$ (g) $xy'z + xy'w'$

17. (a) $(y + xz + x'z')'$ (b) $(x' + yz' + zw' + y'z)'$

18.

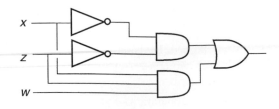

19. Sample answer for LED 1 only: $x + z + yw + y'w'$

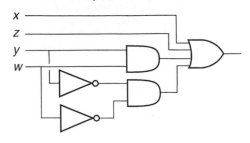

20. Units: $xy' + x'y$; twos: xy

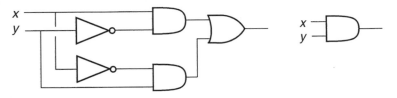

21. Sum: $xyz + xy'z' + x'y'z + x'yz'$; carry: $xz + xy + yz$

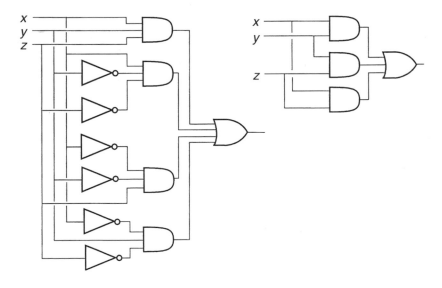

22. (a)

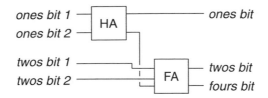

(b)

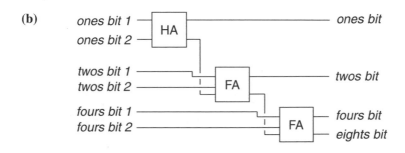

23. AND:

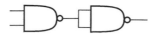

OR:

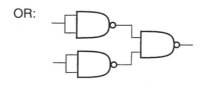

NOT:

CHAPTER 9

1. {tea, coffee} × {full milk, reduced milk, no milk} × {sugar, no sugar}

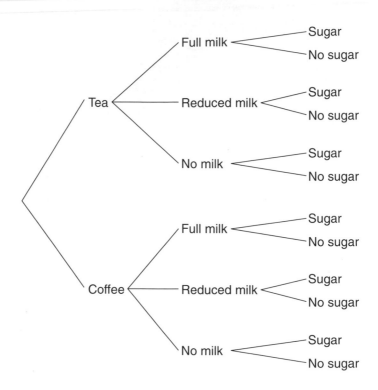

2. 17576000

3. (a) 8000000   **(b)** 4251528   **(c)** 3748472

4. 6100

5. (a) 4096   **(b)** 768

6. 1828641201920

7. 1 hour 48 minutes.

8. (a) Approximately 1 hour.   **(b)** Approximately 41 years.

9. (a) 151200   **(b)** 19958400   **(c)** 1287   **(d)** 1365

10. 3575880

11. 39916800

12. (a) 4368   **(b)** 1764

13. 14

14. (a) 84   **(b)** 840   **(c)** 47040   **(d)** 94080

17. $n = 6$: 1, 6, 15, 20, 15, 6, 1; $n = 7$: 1, 7, 21, 35, 35, 21, 7, 1

18. 2^n

19. (a) For each element x there is a corresponding 1-element subset $\{x\}$.

(b) To each element x there corresponds the $(n-1)$-element subset containing every element except x.

(c) To each r-element subset X there corresponds the $(n-r)$-element subset containing all the elements that are not in X.

Chapter 10

1. (a) Vertices: nodes; edges: electronic components (resistors, etc.); loops: no meaning; parallel edges: components in parallel; directed edges could be used in a DC circuit to indicate current flow.

(b) Vertices: atoms; edges: chemical bonds; loops: no meaning; parallel edges: double or triple bonds; directed edges: no meaning.

(c) Vertices: people; edges: two vertices are adjacent if the corresponding people are friends; loops: no meaning; parallel edges: no meaning; directed edges: no meaning (unless you believe in one-way friendships!).

(d) Vertices: company employees; edges: two vertices are adjacent if one person has managerial authority over the other; loops: no meaning; parallel edges: no meaning; directed edges: used to show the direction of authority.

(e) Vertices: program modules; edges: two vertices are adjacent if one module calls another; loops: used if a module calls itself recursively; parallel edges: could perhaps be used if a module contains more than one calling statement to another module; directed edges: from the calling module to the module being called.

2.

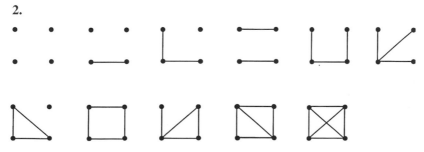

(Check carefully that each graph you have obtained is isomorphic to exactly one of the graphs shown in the figure.)

4. (a) No such graph: the number of vertices with odd degree is odd.

(b)

6. $\dfrac{n(n-1)}{2} - m$

7. (a)

	A	B	C	D
A	0	1	0	2
B	1	1	1	0
C	0	1	1	1
D	2	0	1	0

(b)

	A	B	C	D	E
A	1	0	0	1	0
B	0	0	1	0	1
C	0	1	0	1	0
D	1	0	1	1	1
E	0	1	0	1	0

8. (a) **(b)**

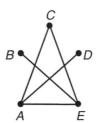

9. (a)

	A	B	C	D
A	0			
B	1	1		
C	0	1	1	
D	2	0	1	0

(b)

	A	B	C	D	E
A	1				
B	0	0			
C	0	1	0		
D	1	0	1	1	
E	0	1	0	1	0

11. (a) $f(A) = B, f(B) = A, f(C) = E, f(D) = D, f(E) = C$
  (b) $f(A) = E, f(B) = F, f(C) = D, f(D) = A, f(E) = C, f(F) = B$

12. (a) The second graph has a vertex with degree 4, while the first does not.
  (b) The second graph is connected, while the first is not.

13. (a)

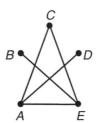

Isomorphism: $f(A) = D, f(B) = A, f(C) = C, f(D) = E, f(E) = B$
  (b) No.

15. (a) Semi-Eulerian. Eulerian path: *cabhjidefg*
   (b) Eulerian. Eulerian circuit: *abceijklgdhf*
   (c) Neither Eulerian nor semi-Eulerian.
   (d) Eulerian. Eulerian circuit: *aikcebhjgfd*

17. (a) *abhjic*
   (b) No Hamiltonian circuit, since it is impossible to visit *A* and *F* without visiting *E* and *G* twice.
   (c) *abdeij*   **(d)** *cebijd*

CHAPTER 11

1. (a) *a, bce, cd, bde*   **(b)** *aed, ef, bcf, afd, bce, abcd*

2.

3. (a)

   (b) No such tree: a tree with 8 vertices must have 7 edges, but the sum of the degrees of the vertices is 16, which does not equal twice the number of edges.

4. 15

5. (a)

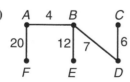

   (b)

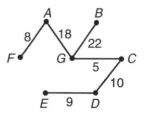

6. (a)

	A	*B*	*C*	*D*	*E*	*F*
A	0	4	∞	∞	15	20
B	4	0	10	7	12	∞
C	∞	10	0	6	∞	∞
D	∞	7	6	0	25	∞
E	15	12	∞	25	0	22
F	20	∞	∞	∞	22	0

(b)

	A	B	C	D	E	F	G
A	0	28	∞	∞	∞	8	18
B	28	0	30	∞	∞	∞	22
C	∞	30	0	10	∞	∞	5
D	∞	∞	10	0	9	∞	12
E	∞	∞	∞	9	0	20	16
F	8	∞	∞	∞	20	0	24
G	18	22	5	12	16	24	0

7. Build the links *AD*, *AE*, *AF*, *BD* and *CF*, with a total cost of 295 cost units.

8. (a) Path *CBAF* with length 34   **(b)** Path *AGC* with length 23

9. (a)   **(b)**   **(c)**

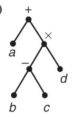
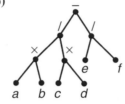
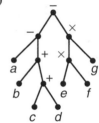

10. (a) $+ a \times - b c d$ **(b)** $- / \times a b \times c d / e f$
    (c) $- - a + b + c d \times \times e f g$

11. (a) $a b c - d \times +$ **(b)** $a b \times c d \times / e f / -$
    (c) $a b c d + + - e f \times g \times -$

12. (a) $((a + b) - c) + (d \times e)$ **(b)** $(a \times (b \times c)) / (((d + e) - f) - g)$

13. (b) 00011011 11110110 01011111 10101101 11110001
    10100011 11000110 11111101 10010

  (c) Letters that occur with higher frequency in English (such as E) can be given shorter codes. This results in coded messages that are shorter on average than if a fixed number of bits is used for each letter.

CHAPTER 12

3.

1. Input n
2. *number_of_primes* ← 0
3. **For** $i = 2$ to n **do**
    3.1. $j \leftarrow 1$
    3.2. **While** $j \leq$ *number_of_primes* and $p_j \nmid i$ **do**
        3.2.1. $j \leftarrow j + 1$
    3.3. **If** $j =$ *number_of_primes* + 1 **then**
        3.3.1. *number_of_primes* ← *number_of_primes* + 1
        3.3.2. $p_{number\_of\_primes} \leftarrow i$
4. Output $p_1, \ldots, p_{number\_of\_primes}$

4. (a) 2, 3, 5, 7, 11, 13, 17, 19, 23, 29, 31, 37, 41, 43, 47

(b)

   1. Input n

   2. **For** $i = 2$ **to** n **do**

      2.1. *prime*$(i) \leftarrow$ true

   3. **For** $i = 2$ **to** n **do**

      3.1. **If** *prime*$(i) =$ true **then**

          3.1.1. **For** $j = 2$ **to** n div i **do**

               3.1.1.1. *prime*$(ij) \leftarrow$ false

   4. **For** $i = 2$ **to** n **do**

      4.1. **If** *prime*$(i) =$ true **then**

          4.1.1. Output i

6. (a) gcd $= 11$, lcm $= 15444$ **(b)** gcd $= 8$, lcm $= 197912$

   (c) gcd $= 1$, lcm $= 3211860$

7. (a) $\dfrac{279}{53}$ **(b)** $\dfrac{61}{97}$

8. (a) $x = 699 + 5726n$, $y = -558 - 4571n$, $n \in \mathbf{J}$

   (b) No solution: $11 \mid 2783$ and $11 \mid 2046$ but $11 \nmid 10$

   (c) $x = 151 + 2091n$, $y = -289 - 4002n$, $n \in \mathbf{J}$

9. (a) $x = 200 + 5n$, $y = -300 - 8n$, $n \in \mathbf{J}$

   (b) $x = 10$, $y = 4$; $x = 5$, $y = 12$

11. $E(0) = \{\ldots, -10, -5, 0, 5, 10, \ldots\} = \{5n: n \in \mathbf{J}\}$;

   $E(1) = \{\ldots, -9, -4, 1, 6, 11, \ldots\} = \{5n + 1: n \in \mathbf{J}\}$;

   $E(2) = \{\ldots, -8, -3, 2, 7, 12, \ldots\} = \{5n + 2: n \in \mathbf{J}\}$;

   $E(3) = \{\ldots, -7, -2, 3, 8, 13, \ldots\} = \{5n + 3: n \in \mathbf{J}\}$;

   $E(4) = \{\ldots, -6, -1, 4, 9, 14, \ldots\} = \{5n + 4: n \in \mathbf{J}\}$

12.

$+$	$E(0)$	$E(1)$	$E(2)$	$E(3)$	$E(4)$
$E(0)$	$E(0)$	$E(1)$	$E(2)$	$E(3)$	$E(4)$
$E(1)$	$E(1)$	$E(2)$	$E(3)$	$E(4)$	$E(0)$
$E(2)$	$E(2)$	$E(3)$	$E(4)$	$E(0)$	$E(1)$
$E(3)$	$E(3)$	$E(4)$	$E(0)$	$E(1)$	$E(2)$
$E(4)$	$E(4)$	$E(0)$	$E(1)$	$E(2)$	$E(3)$

$\times$	$E(0)$	$E(1)$	$E(2)$	$E(3)$	$E(4)$
$E(0)$	$E(0)$	$E(0)$	$E(0)$	$E(0)$	$E(0)$
$E(1)$	$E(0)$	$E(1)$	$E(2)$	$E(3)$	$E(4)$
$E(2)$	$E(0)$	$E(2)$	$E(4)$	$E(1)$	$E(3)$
$E(3)$	$E(0)$	$E(3)$	$E(1)$	$E(4)$	$E(2)$
$E(4)$	$E(0)$	$E(4)$	$E(3)$	$E(2)$	$E(1)$

14. (a) $E(1)$, $E(2)$, $E(4)$, $E(5)$, $E(7)$, $E(8)$ **(b)** $E(1)$, $E(3)$, $E(7)$, $E(9)$

15. (a) $x \equiv 17 \bmod 19$ **(b)** $x \equiv 18 \bmod 27$

16. 1, 2, 9, 10, 1, … (repeats from this point on).

17. 1, 4, 15, 2, 13, 0, 11, 14, 9, 12, 7, 10, 5, 8, 3, 6, 1, … (repeats from this point on).

20. Sample mean: 0.514423; sample variance: 0.095786

CHAPTER 13

1. Dominant operation: multiplication in Step 3.2; time complexity: $n - 1$

2. (This answer refers to the algorithm given as the answer to Chapter 1, Exercise 6.) Dominant operation: comparison in Step 4.1; time complexity: $n - 1$

3.

Step	d	i	b_i	x	answer	Output
1	–	–	–	2	–	–
2	4	–	–	2	–	–
3	4	–	–	2	–	–
3.1	4	–	–	2	2	–
4	4	2	1	2	2	–
4.1	4	2	1	4	2	–
4.2	4	2	1	4	2	–
4.2.1	4	2	1	4	8	–
4	4	3	0	4	8	–
4.1	4	3	0	16	8	–
4.2	4	3	0	16	8	–
4	4	4	1	16	8	–
4.1	4	4	1	256	8	–
4.2	4	4	1	256	8	–
4.2.1	4	4	1	256	2048	–
5	4	4	1	256	2048	2048

4. (a) 6   **(b)** 5

5. B is faster.

8. $10n + 13$, $n - 4$, $n \sqrt{n}$, $n^2 + n \lg n$, $n^2 \lg n$, $n^3 + 4n$, 3×2^n; the first two complexities in the sequence are equivalent.

9.

Step	i	x_1	x_2	x_3	insert	lower_limit	upper_limit	j	k	Output
1	–	8	5	7	–	–	–	–	–	–
2	2	8	5	7	–	–	–	–	–	–
2.1	2	8	5	7	5	–	–	–	–	–
2.2	2	8	5	7	5	1	2	–	–	–
2.3.1	2	8	5	7	5	1	2	1	–	–
2.3.2.1	2	8	5	7	5	1	1	1	–	–
2.4	2	8	5	7	5	1	1	1	1	–
2.4.1	2	8	8	7	5	1	1	1	1	–
2.5	2	5	8	7	5	1	1	1	1	–
2	3	5	8	7	5	1	1	1	1	–
2.1	3	5	8	7	7	1	1	1	1	–
2.2	3	5	8	7	7	1	3	1	1	–
2.3.1	3	5	8	7	7	1	3	2	1	–
2.3.2.1	3	5	8	7	7	1	2	2	1	–
2.3.1	3	5	8	7	7	1	2	1	1	–
2.3.2.2	3	5	8	7	7	2	2	1	1	–
2.4	3	5	8	7	7	2	2	1	1	–
2.4.1	3	5	8	8	7	2	2	1	1	–
2.5	3	5	7	8	7	2	2	1	1	–
2	3	5	7	8	7	2	2	1	1	–
3	3	5	7	8	7	2	2	1	1	5, 7, 8

12. (a) $O(mn)$ **(b)** $O(mn^2)$

13. Not tractable: a set with n elements has 2^n subsets, so any such algorithm would have time complexity at least $O(2^n)$.

INDEX